Size
Really Does
Matter
The Nanotechnology Revolution

Size
Really Does
Matter
The Nanotechnology Revolution

Colm Durkan

University of Cambridge, UK

 **World Scientific**

NEW JERSEY · LONDON · SINGAPORE · BEIJING · SHANGHAI · HONG KONG · TAIPEI · CHENNAI · TOKYO

Published by

World Scientific Publishing Europe Ltd.

57 Shelton Street, Covent Garden, London WC2H 9HE

Head office: 5 Toh Tuck Link, Singapore 596224

USA office: 27 Warren Street, Suite 401-402, Hackensack, NJ 07601

Library of Congress Cataloging-in-Publication Data

Names: Durkan, Colm, author.

Title: Size really does matter : the nanotechnology revolution / by Colm Durkan, University of Cambridge, UK.

Description: New Jersey : World Scientific, 2019. | Includes index.

Identifiers: LCCN 2018049287 | ISBN 9781786346612 (hc : alk. paper)

Subjects: LCSH: Nanotechnology. | Nanostructured materials.

Classification: LCC T174.7 .D87 2019 | DDC 620/.5--dc23

LC record available at https://lccn.loc.gov/2018049287

British Library Cataloguing-in-Publication Data

A catalogue record for this book is available from the British Library.

First published 2019 (Hardcover)

Reprinted 2019 (in paperback edition)

ISBN 978-1-78634-797-8 (pbk)

For any available supplementary material, please visit
https://www.worldscientific.com/worldscibooks/10.1142/Q0197#t=suppl

Desk Editors: Vimal Thangavel/Jennifer Brough/Shi Ying Koe

Typeset by Stallion Press
Email: enquiries@stallionpress.com

Printed in Singapore

For Mary Rose — for your unquestioning support and love through it all.

Preface

This book has been what you might call a labor of love. I believe passionately that science deserves and needs to be communicated to a wider audience. The world around us is so fantastically complex yet utterly beautiful, with patterns and laws to be seen at the scale all the way from the size of the universe to the size of the fundamental particles. Many wonderful books have been written about this at the size of the universe, but few have managed to convey the wonder at the size of atoms and just a bit bigger. This is what I have set out to do. I hope that I have managed to strike the balance right, at least most of the time. Nanotechnology is at that exciting point where it has evolved from the lab and grown legs of its own and become part of our everyday lives, without our even realizing. The point I want to get across is that this need not be a concern. The fundamental underpinning science to nanotechnology is not that new and we have over 100 years of experience to draw on. At the heart of nanotechnology is the fact that small pieces of matter, and when I mean small, I of course mean nanometer (1/1000,000,000 of a meter)-sized, have very different properties than larger pieces of the same material. It is all about understanding that and harnessing it to our advantage to make, for example, (i) stronger and lighter materials, (ii) better conductors of heat and electricity than conventional metals, (iii) materials whose color changes whenever they change size, and (iv) better, faster, and less harmful ways of detecting and treating cancer. These are but the tip of the iceberg. In this book, I am concentrating on those topics that I personally believe are of most interest to most people. I am also assuming that it will grow over time to include new areas, both in research and in application. My aim is that by reading this book, you will feel more confident to assess any headlines you see that mention anything about nanotechnology — either good or bad, as to how realistic any claims made actually are. This is one of my guiding principles — the general population of many developed countries is far more scientifically literate than their press give them credit for, and people do not want to be patronized. As we improve existing technologies and develop new ones, we need to be ever watchful and aware of the drain on our natural resources. We are at a point in our civilization where power and resource management are at a critical point, especially due to the growth in the internet and in motor car electrification. As more and more of us are living well beyond the age that the human body was designed for, we are encountering new diseases which we need to combat. The same is true of foodstuffs — the drive to produce larger quantities of food and the consequent processing required has had unintended consequences to our health. These are some of the issues that we will touch on in this book, which is really intended as an overview of what nanotechnology is, where it came from and where it is most likely to benefit us in the future. It was a difficult task as I will mention in Chapter 1, as there are few hard numbers to draw on — mostly just underlying principles and concepts. Dip in and out of the chapters, but for the story to make sense in its entirety, first read Chapters 1–3 as they provide the background that is used throughout. Chapter 5 is possibly the hardest to explain — it is about how we actually see things that are nanosized, as after all, they are very small. I firmly believe that the first step toward understanding what something is and how it works is to look at it. This is particularly important in nanotechnology due to the intimate correlation between size, shape, and properties.

This book would not have been possible without the unquestioning support of Mary Rose, to whom I dedicate it. Thanks also go to Rosie (who urged me to keep writing and has sweet delusions of this becoming a bestseller), Ben (who said there were far too many commas), Joe Durkan, Valerie Holt, Greg Durant, and any others who looked at the various drafts and tactfully told me that what I had written was either good or rubbish. Thanks also to the family dog, Spencer, for dragging me out for lots of long walks where I was able to escape the chaos and formulate what I wanted to impart in this book. I am also immensely grateful to my editor, Lance Sucharov at World Scientific, for his input and for believing in me and to Jennifer Brough for helping with the final tweaks needed to get this to press. I have presented much of this material to a range of people from schoolchildren to retired submarine engineers and artists, who have all found something in it to interest them, and to me, that's what it's all about. Science is the first step toward technology which is an enabler in our lives. Science for science's sake has its place, but technology is the true test of which science is actually useful, at least for now. So, with that, I hope you find something in these pages to interest you, and happy reading!

Colm Durkan
Cambridge

May 2018

About the Author

Dr Colm Durkan is the founder and Head of the Applied Nanoscience and Nanoscale Engineering research group at the Nanoscience Centre of the University of Cambridge. He is also a fellow of Girton College, Cambridge. He obtained his primary degree and PhD in Physics from Trinity College Dublin. After spending a postdoctoral stint at Konstanz University in Germany, he came to work in Cambridge in 1997 where he has been ever since. He has been a university lecturer with full tenure since July 2000 and a Reader in Nanoscale Engineering since October 2010. At the university, he has developed and taught courses on nanotechnology, quantum mechanics and electronics. He was elected to a fellowship of the Institute of Physics in 2009 and the Institute of Engineering & Technology in 2014 for his work in nanotechnology.

Colm's research interests lie in understanding how and why the properties of materials change dramatically when those materials are of nanometer dimensions and how to make practical use of that, and in the properties of surfaces. He has led industrial collaborations with BP, Unilever, Samsung and Nokia, and loves nothing more than exploring new phenomena in the lab. Colm lives in Cambridge with his family.

Contents

Chapter 1

Introduction

We exist in a universe governed by the rules of quantum mechanics. Rules that we unfortunately do not understand and may never be able to. At the turn of the 20th century, the development of quantum theory led to the first steps toward nanotechnology. This theory is all about the physical rules that fantastically small things are seen to obey. These rules are different from what we encounter in our everyday lives as they relate to effects that are too small to see directly. They do, however, become important once we are dealing with nanometer-sized things, which is what nanotechnology is all about.

It is at the boundary of existing knowledge where the most useful and significant discoveries are often made, and this is what makes nanotechnology so exciting — it is concerned with the nature of *things* that are a few nanometers across, and encompasses physics, chemistry, biology, medicine, and engineering. In fact, the Nobel Prize for chemistry in 2016 was awarded to three nanotechnologists for their pioneering work in creating molecular machines, which we will explore in Chapter 4.

The word *Nano* comes from the Greek "Nanos" meaning "dwarf", and in the scientific context the prefix *nano* is used to indicate one billionth (a billion being a thousand million). One *nano*meter, abbreviated as *nm* is therefore one billionth of a meter. Nanotechnology is defined as the ability to both create and study objects with at least one dimension in the size range 1–100 nm. The diameter of a strand of DNA — our genetic blueprint is 2 nm, and the spacing between atoms in a typical solid material is around a fifth of a nanometer. In much the same way as getting to grips with the size of the universe is challenging for us — we know that it is at least 46 billion light years across, perhaps even infinite (we don't actually know!); the size of things down at the atomic scale is so small that we find it equally hard to envisage. We can explore the observable universe using telescopes,

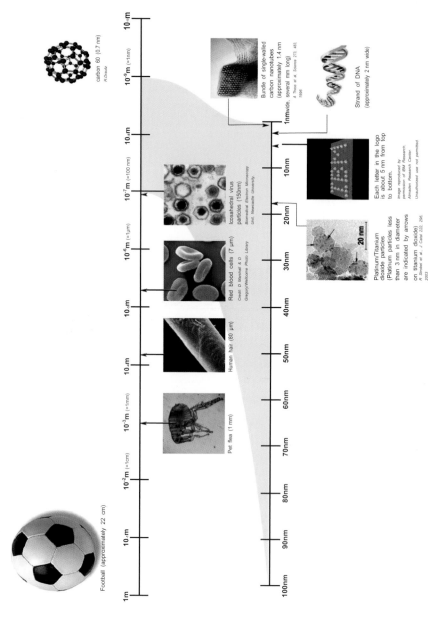

Figure 1.   Length scale showing the nanometer in context. The length scale at the top ranges from 1 m to 10^{-10} m, and illustrates the size of a football compared to a carbon 60 (C60) molecule, also known as a buckyball. For comparison, the world is approximately 100 million times larger than a football, which is in turn 100 million times larger than a buckyball. The section from 10^{-7} m (100 nm) to 10^{-9} m (1 nm) is expanded below. The lengthscale of interest for nanoscience and nanotechnologies is from 100 nm down to the atomic scale — approximately 0.2 nm.

Source: Relative scales schematic from the Royal Society report "Nanoscience & nanotechnologies: opportunities and uncertainties", 2004.

and similarly we can explore the nanometer world using microscopes. This is a key element of nanotechnology and nanoscience as we will see, particularly in Chapter 5. A useful guide as to the relative size of everyday objects from 1 m down to 0.1 nm is shown in Figure 1 which was produced by the Royal Society in 2004.

While we are thinking about ridiculously large numbers and small things, an example is to consider roughly how many atoms are in a typical person, say, me. We will do this by working out the ratio of my volume to that of a single atom, which should equal the number of atoms in me. I weigh approximately 73 kg, and given that the density of the average human is 0.98 times that of water (which is 1000 kg per cu. m at 20°C), my volume = mass/density = 73 kg/(0.98 × 1000 kg per cu. m) = 0.07 cu. m. The volume occupied per atom in a material is around 10^{-29} cu. m, meaning I contain approximately $0.07/10^{-29}$, or 7×10^{27} atoms, which is 7 billion billion billion atoms (Figure 2). An enormous number to be sure, but this is nothing compared to the number of bacteria on the planet, which is estimated at around a thousand times more, or 5×10^{30}. If I were to take each of those atoms in me and spread them out in a line with their typical spacing of around 0.2 nm, then that line would be 1.4×10^{18} m long, which is 149 light years, almost long enough to get us to the Hyades star cluster in the constellation Taurus.

7,000,000,000,000,0 00,000,000,000,000

Figure 2.   The approximate number of atoms in my body: 7×10^{27}.

I have been working in nanotechnology since the mid-1990s, and in the short time from then until now, have seen a renaissance in the way we do science and in how it is communicated to a wider audience. I have seen scientists treated simultaneously with respect, contempt, and distrust, and have had many fascinating conversations with all sorts of people who really just want to know more. I have met those who believe science is a mere vanity and utterly useless and those who believe we have a duty to society to share our knowledge and dig deeper. Why does mankind strive to climb our planet's highest mountains, to dive to the ultimate depths of the sea bed, or to travel out to space? Do we actually learn anything about this apart from how fragile human life is, how tenacious life in general is, and how small we are in the grand scheme of things? These are difficult questions to answer but they do need to be asked. All I can say is that it is part of the human condition to question and want to know more about our surroundings, both near and far. Without this thirst for knowledge, I have no doubt we would still be swinging out of trees in the jungle. As an academic, I may of course be viewed as part of the establishment, but I can see the merit of all sides of this argument and that it is not so simple — some research is almost certainly never going to lead to any useful outcomes (This is not always known at the time of course), and yes, we do have a duty to wider society to use our funding responsibly and communicate our findings in as accessible a manner as possible.

It was at the age of around 12 that I became interested in science when I was given a copy of a new magazine called *Science Now*, a Marshall Cavendish publication that ran weekly. There were articles on sharks, astronomy, aeronautics, space exploration, medical advances, and all manner of things that were just of general interest, but they got me wondering about how things work. Rather than go down the path of engineering (which I now teach), I chose to pursue physics, as to be honest, as a teenager growing up in Ireland, I thought engineers were the people who fixed washing machines and boilers. I still meet those who think the same and it's no surprise, as the word is constantly misused. On the continent, engineers and scientists are placed on the same level socially as doctors and lawyers, respected and considered to be essential cogs in the machinery of a well-functioning society, whereas in the British Isles, we seem to have lost this sentiment some what, so it is time to earn it back. The world is not full of mad scientists sneaking around trying to make new discoveries at all costs, not caring about ethics or the effect on the world, apart from the unscrupulous few. We are just normal people who have a burning curiosity about how things work and want to find out more with the overarching aim of wishing to use that knowledge for the greater good. The ultimate reward for any scientist is to do something which is *useful* to people in their everyday lives. This can be seen by others as quirkiness but I would rather label it as tenacity and an ability to focus. In my case, the initial interest in physics led me down a wandering path where my work straddles physics, chemistry, materials science, engineering, and biology, with a single common thread running through it all. I am fascinated by microscopic things, i.e. things too small to see with our own eyes. In fact, I am most interested in things that are even smaller than just microscopic, and we could describe as *nanoscopic*. By virtue of their size, things this small have revolutionary properties and my job in writing this is to convince you that not only is this useful and in fact is already being used in a vast number of cases, but it will also change the way we view the world around us. The world and in fact the universe is utterly beautiful in its complexity, and this is only enhanced as our understanding deepens. The fact that we have developed sufficient insights in the past few millennia to be able to explain everyday things is a marvel. A common mistake of the less scientifically literate is to believe that the world must be very dull to scientists as there is less mystery in what we observe. We understand why flowers are the color they are and how this has evolved to attract the right kind of insects or pollinators, how the tides work, how the stars and planets move across the sky, how the sun works, how to predict the weather, etc. The fact that there are patterns in nature that we can describe using models that *make sense* is where the true beauty of nature lies. Some of the subtler intricacies of nature are only revealed at nanometer scales, hence my assertion that by finding out more about it, we will start to understand more about the world around us and look at it through a different lens. So, in a nutshell, this is a book about nanotechnology and how it affects our lives now and how it will facilitate new developments in the future.

Nanotechnology is everywhere. That sounds a bit dramatic to say the least. It is however, true. I had better expand on what I mean by *everywhere* so we don't get carried away before we even start. I really mean that nanotechnology has in some way touched on many aspects of our daily lives, in mostly good ways, although not always so. We live in an age where many things are transient and immediate, including knowledge, with the result that the subtleties of nature can easily be overlooked unless we take a step back and look closely. At the same time, we are making great strides in technology, particularly artificial intelligence (AI), machine learning, and the internet of things. We have harnessed the power of the atom both for great good (atomic energy) and great harm (nuclear weapons). Many of the technologies of the modern age that we now take entirely for granted have been made possible through advances in manufacturing of electronic devices and components and the discovery of new materials — from composites to graphene. Out of these early developments, nanotechnology was born and has infiltrated all industry sectors, running in the background. This may sound sinister, but actually nanotechnology is in many ways obvious as we will see — it is an approach to doing things that makes optimum use of the strange and wonderful properties of materials when they are made very, very small.

Continuous developments in science and technology have inexorably led us to the point where we have kept making things smaller and smaller. This is particularly so in the electronics industry where individual circuit elements have become small enough that the size of individual atoms is noticeable and starts to matter. This is also the size or *lengthscale* where the quantum nature of our world is revealed, and that is no coincidence. As we will see later, quantum mechanics may be entirely freaky and weird, but it is also unquestionably correct. We therefore need to get to grips with what it all means as we are in the middle of a step change in our technological capability as a result of it.

Now we can start to delve a little deeper in our quest to find out what nano-technology is. If I give you a hint — the *nano* bit is a measure of size — the *nanometer*. Nanotechnology is all about the science and technology of things whose size we measure in nanometers, which is 1 billionth of a meter. You have probably heard or read *something* about nanotechnology, ranging from the prepos-terous to the downright terrifying, and possibly loosely based on some half-truths or dare I say, *alternative facts*. For a while in the late 1990s and early 2000s, nanotechnology was touted as the next best thing that would revolutionize our world, and then the promise turned to suspicion akin to that encountered by the field of genetic modification in crops. Thankfully, we seem to be out of the woods and nanotechnology is again being seen as inevitable and something to be consid-ered and harnessed rather than feared and fought.

Nanotechnology is a form of technology that has become pervasive. It already affects our daily lives in many sectors including but not limited to cosmetics, packaging, construction, healthcare, art materials, defense, oil and gas, energy storage, and clothing. This is captured in Figure 3.

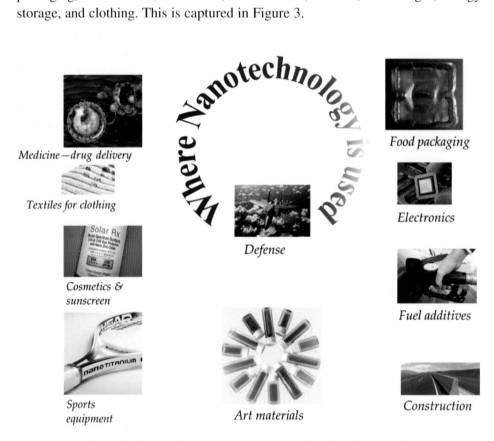

Figure 3.   The application areas where nanotechnology is most used.

The aim of this book is to explore this in a systematic way. In the course of this, I will dispel some myths that have arisen. One example of this has got to be one of the most annoying misconceptions about nanotechnology that I feel duty bound to dispel the myth entirely. There have been many very nice artists' renditions in a variety of journals and newspapers over the last 20 years depicting tiny robots moving around in our bloodstream repairing damaged cells and tissues, and this idea was originally touted by nanotechnologists seeking funding for their research. While nanotechnology is and will continue to lead to major breakthroughs in medicine that are actually useful, it is not sensible to think of nanorobots going around our bodies repairing damaged cells. This is a compelling idea and there's no doubt, Hollywood has used this concept time and time again, in otherwise quite cool and fun movies. But it's just wrong! It has nothing to do with nanotechnology. I'm not saying it's impossible, but I am saying that it is highly unlikely — there are much better ways of healing where we use nature or medicines rather than machines. Let's not fill our and our children's heads with stuff that's just complete twaddle and try to get it right!

I have written this book as a story — how we came to realize that the world around us is far more complex than it appears, and that it is in fact really quite beautiful in the patterns that appear in the laws of nature, with an eye on all things nano.

I will focus on the development of nanotechnology with the emphasis on things that have recently been discovered and are either under development in research labs or employed in industry, and therefore in everyday life. I am not so interested in speculating about things that may or may not happen, no matter how exciting they may sound. Nanotechnology as a topic is vast, so in order for this book to not be thousands of pages long and utterly disjointed, I will concentrate on those aspects that I consider most important.

My own research is firmly rooted in a basic-principles approach which is applied to fields as diverse as quantum devices, understanding fouling in oil-exposed pipes, the effect of various common treatments on the molecular-scale properties of hair and teeth, and a whole host of things to do with how molecules of all kinds interact with surfaces of all kinds, as this is a surprisingly common problem. The inevitable price I will pay for this cherry-picking approach is that I will leave things out and somebody somewhere will say "hey, why didn't you mention space travel or batteries or invisibility or …". However, the purpose of this book is that once you have read it, if you come across some headline or other that talks about the latest breakthrough in nanotechnology, you will have gained the tools to analyze it critically and make up your own mind. I will not draw on hard numbers for many things as they simply don't exist. The topic is highly conceptual, but there are lots of everyday examples that we can look at to illustrate all of the relevant points. Nanotechnology is about making and using nanometer-sized things, which means *very, very small things just a bit bigger than single atoms*. The size or *lengthscale* of objects are absolute but our perception of them is subjective from the point of view that while there is no doubt 1 nm is pretty small, it is *enormous* compared to the suspected size of an electron, proton, neutron, or Higgs Boson, which are at least a million times smaller. For people working in particle physics, a nanometer is huge, whereas to people working in cosmology, the size of our solar system is tiny, never mind a nanometer!

Why then is the nanometer an important lengthscale — who cares? This is what we will look at in this book, at the fact that the properties of matter are to a large extent determined by phenomena at the nanometer scale. It is no coincidence that some of the fundamental lengthscales of physics and chemistry happen to be in nanometers. What I mean by lengthscales is that once objects are within a range of sizes, certain effects start to be noticed. For example, the wavelength of waves on the surface of the sea is of order 10 m. If we are on a large boat, for instance, the new Queen Elizabeth aircraft carrier which is 280 m long, then we will not notice those waves at all. On the other hand, if we were in a small boat, say 5 m long, then we will be thrown all over the place by those waves. Another example is the size of an atom — around 1/10 of a nanometer. If we were to somehow have an atom thrown at us, we would not even notice it. However, if we were to throw that atom at another atom, they would both know about it. It is all to do with the relative sizes of things or characteristic distances over which certain effects are noticed. It turns out that there are a few of these lengthscales in physics, and they happen to be in the range of nanometers, and that is why nanotechnology even exists. An example of something that has dimensions in nanometers is DNA — our genetic blueprint consisting of two long molecular strings, collectively called a double helix. This has an average diameter of 2 nm (Figure 4). Size therefore is entirely relative and really *does* matter.

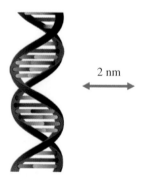

2 nm

Figure 4.   DNA — without a doubt the most important nanostructure!

As it turns out, because of this issue to do with lengthscales, things that are small enough to talk about their size in terms of nanometers have very different characteristics and properties (in terms of their strength, color, chemical reactivity, and a whole host of other things) than things that are bigger. Nanotechnology is all about making use of this to either do something new or to improve something we already do, be that in manufacturing processes, healthcare, or any of the other sectors I have mentioned above that we will look into. In this book, we will explore these ideas in detail, and look at how nanotechnology is being employed in a wide range of applications and assess its potential importance to society and our daily lives. Nanotechnology has exploded on the scene over the last few years as we have only recently become rather good at making such small things and then, most importantly — making use of them.

Having spoken to many non-scientists about this over the years, I have come to the conclusion that if I start this book with all of the whizzbang things that nanotechnology can do and then try to fill in the gaps and background then you will lose interest too soon and this book will end up gathering dust. Or maybe be used for firewood or to prop up a wonky desk. Either way, not being read! On the other hand, if I slowly work through the history and build it up bit-by-bit to a grand finale, then you will get impatient and complain that I took too long getting to the point. Instead, I have decided to give a flavor of what nanotechnology is all about in the introduction and then have a look at the background before starting to look at different sectors, namely materials, computer technology, and finally medicine. Each chapter in principle is stand-alone, but it is best to read the whole book as the background is particularly important when trying to see where things are heading, and also, *it's all interlinked*. Armed with this knowledge, I will start with a brief overview of where we are now. I will then delve back to the beginnings of the field so that we can get an idea of where we are heading, in order that we can make informed choices and not have to rely on alarmist and uninformed headlines.

Today, nanotechnology is integral to the multi-billion dollar industries already mentioned — healthcare, food, clothing, oil & gas, electronics, cosmetics, automotives, sporting goods, art, construction, and defense. New treatments for some forms of cancer are being developed that use nanotechnology to target tumors and reduce the side-effects of chemotherapy. Some clothing has a nanoparticle-containing coating on it to render it waterproof. Microprocessors that are at the heart of computers have literally billions of transistors that are just a few tens of nanometers across, and by the end of the 2020s, they will be less than 10 nm across. Many of the cosmetic products that we use every day contain the active ingredients in nanoparticle form. The automotive industry utilizes nanometer-sized additives to fuel in order to enhance performance and increase engine lifetime. Sporting equipment is often made using alloys or composites containing nanoparticles or nanofibers. In art, the colors used for painting contain nanoparticles as pigments. Nanometer-thin layers are added to glass to make it self-cleaning, and new composite materials are widely used in the automotive, sporting, and defense industries for their enhanced strength and lightness. On the minus side, airborne nanoparticles produced by internal combustion engines in cars are in part responsible for the rise in respiratory illness and disease including lung cancer and have led to significant changes in legislation regarding the future of the automotive industry.

A web search of "Nanotechnology" will return over 7 million entries, whereas the underpinning science or "Nanoscience" only returns around 1.5 million entries. There are several variations, so perhaps a more meaningful search is just "Nano", which appears 59 million times. To put it in context, web searches for "Physics", "Chemistry" and "Biology" return 92 million, 93 million, and 79 million entries, respectively, whereas "Medicine" and "Engineering" return 202 million and 349 million, respectively. See the pattern? More established fields have more entries, and those with the most entries, which I am taking as a rough measure of their perceived importance to the average person are those that have a direct impact on our daily lives, i.e. engineering and medicine. A word of caution though — this should all be taken with a grain of salt as "dog" appears 261 million times, "cat" 379 million times, and "Facebook" over 5.2 billion times! In this book, we will look at the above issues in more detail. First, however, I would like to briefly explore how it all began. A word about my writing style: I am a physicist, so I have put in numbers for things whenever I can, but the field of nanotechnology is so ridiculously vast, that there are topics for which there are no reliable numbers, so we will instead look at representative examples. I have not, however, put in many formulae, as they are not necessary in a descriptive book such as this, but believe me when I say that this is painful for me! I am a die-hard believer in using maths in physics as to me, maths is the true language of physics and enables us to go far beyond mere descriptions (Otherwise known as "pub physics"), and as a parent of school-age children, I find it hard to spot where the actual physics is in the syllabus, so do not wish to add to this dumbing-down. Without the deeper understanding imparted through the use of maths (which in turn is made possible through appropriate insight), physics appears as a set of disjointed facts, which is why I believe so many schoolchildren find it so difficult and therefore don't like it which is utterly tragic. This is particularly so with concepts in electricity. This is completely avoidable, just like dental decay — my better half is a dentist, so I hear this a lot. And don't get to eat many sweets. And brush my teeth regularly. You get the picture…

In the Beginning...

In order to appreciate the current state of our scientific understanding of the world around us and how many paths inexorably led to nanotechnology, we will look at how our understanding of science has evolved through the ages.

The early 20th century was a tumultuous time for physics in particular with the realization that the universe is far more complicated than anyone had ever realized, leading to the necessary development of relativity and quantum mechanics. Towards the end of the 19th century, physics had become an altogether stale discipline, and an eminent scientist[1] remarked around 1890 that "there is nothing new to be discovered in Physics now. All that remains is more and more precise measurement". What a blooper that turned out to be. It was around this time that biology was starting to grow in stature as a scientific endeavor of relevance, with Darwin's theory of Natural Selection and Mendel's ideas on genetics becoming more accepted. At that time, it was widely felt amongst the scientific community that physics had had its day. Many of us now accept the overwhelming scientific evidence for evolution. At the time that Darwin published *On the origin of species*, the book sold out rapidly on its first print run in 1859, and was met with approval by and large, apart from by the church, who either disregarded it, saw it as heresy or used it as a further proof of God's design of the world around us, and that mindset persisted for some time. The disapproval of Darwin's ideas by the non-scientifically minded led to the sort of derision that is clear from the drawing on the next page. In fact, although the Catholic Church has accepted since 1950 that some of its congregation believe in evolution, it was only in late 2014 that the pope officially stated that evolution and the big bang are scientific facts, but argued that they do not rule out the existence of a divine creator. There are many such examples where western religion has been at odds with science, but in the end, science always wins as you can't argue with facts. This is not intended as a statement against religion, merely a cautionary note about where it probably should not be invoked.

Leaving out any religious bias, the certifiable scientific evidence for evolution was there in black and white for all to see, so a few years after publication of his book, in 1864, Darwin was honored by the Royal Society who awarded him the Copley Medal, their highest honor. There are few cases where revolutionary ideas are accepted so rapidly, as there is a general distrust of new concepts, which take time to be verified and to sink in, and eventually become accepted as obvious.

---

[1] This is often attributed to Lord Kelvin, but was actually stated by Albert Michelson, one of the first people to accurately measure the speed of light, which is just under 300,000 km per sec.

An example is the aforementioned work of Mendel — by performing experiments on the breeding of peas, he noticed that there were seven traits (Plant height, seed shape, and color, pod shape and color, and flower position and color) that seemed to be inherited. He grew just under 30,000 pea plants over a 7-year period and looked at the similarities in the appearance of the pea seeds and the resulting plants, which led to his observations that the above seven traits are passed from one generation to the next. He presented his findings in 1865, which were largely ignored as people either weren't that interested in peas or couldn't see that this was a universal concept that could be applied to *all living things*. It was only after his death (in 1884) that his ideas were rediscovered within the context of Darwin's and started to gain acceptance. His work led to the notion that for traits, such as size, color, leaf shape, etc., to be inherited, then some sort of information must be passed on from one generation to the next (Figure 5). The exact mechanism behind this was unclear, but these basic units of information became known as *genes*. Even today, heated arguments about the deeper nature of this are still raging — is natural selection happening at the level of the gene, the organism, the community, or the species?

Our ability to shape our environment and to prolong and improve the lives of those with chronic illnesses is inevitably going to change the direction of human evolution. Through embryonic screening and with the advent of the capability to create designer babies (nothing to do with nanotechnology!), basically forms of eugenics, many issues deeply challenging our ethical sense are being raised, and

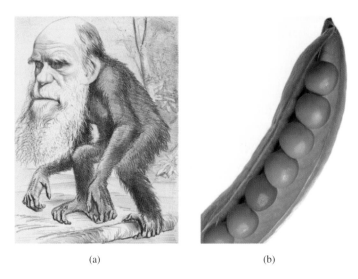

(a)   (b)

Figure 5.   The advent of genetics in the latter half of the 19th century contributed to the shift of interest away from physics. (a) An irreverent sketch of Darwin from 1871; (b) The humble pea was used by Mendel to tease out the principles of genetics.

it is the case that we have the ability to tailor future human populations — an utterly terrifying and unsettling thought. Of course, this is already happening, for example, in Sweden, no children with Down's syndrome are born any more now as all pregnancies are screened for it. How did these things come about? Through research on various diseases and on the human genome, the tools that enable those practices were developed. Science gives us the tools to overcome things that challenge us, and ethical issues are then raised depending on how we then apply that science. I can safely say that nanotechnology does not tread on toes in this way so I can hold my hand up and say I am not concerned about the misuse of anything that I may discover. We know that medicines work better for some individuals than for others and ultimately this is down to genetic differences. We are on the verge of having personalized healthcare where the medicines we take will be matched to our genetic code. A lot needs to happen first though, some of which we will look at in this book.

Against the backdrop of these other scientific developments, general interest in physics was on the decline in the mid-19th century. However, with advances in measurement technology, it became apparent that there were deviations from the expected behavior of many things, something that was first noticed in astronomy. For example, the orbits of the planets were found to be ever so slightly inconsistent with what was expected from Newton's laws, which had been thought to be all we needed to explain how things move and had worked perfectly well for over 200 years since Newton's time.

A second example is that the electrical properties of materials were baffling to explain — we had no idea why some materials were able to carry electric currents, and others not, and the discovery of the electron by J.J. Thomson in 1897 revealed that in fact many of our scientific theories and ideas about the universe around us were mere *approximations*. The quest then began to unravel the secrets of nature and to find out what these previous theories were approximations to, starting with Einstein's work on relativity and quantum mechanics, followed by a host of other breakthroughs mainly in quantum mechanics. In 1906, J.J. Thomson from Cambridge, received the Nobel Prize in Physics for his discovery of the electron, one of the primary constituents of atoms, and the charged particle that electric current is made of, and demonstrated that it was a particle (i.e. a solid object). Then, in 1937, his son, George, won the same award for demonstrating that the electron acts as a wave in line with the predictions of quantum mechanics! That must have led to all sorts of interesting breakfast-table conversations. I can just imagine the heated arguments — "it's a *particle*!" ..."no, dad, it's a *wave*! You never listen!" ..., etc. Having teenage children hones your mind to such frank exchanges... It is also most humbling that I am sitting writing this in my office on J.J. Thomson Avenue in Cambridge at a time when we know that electrons, and in fact, *all* objects behave as both particles *and* waves. Which type of behavior you see depends on the circumstances. If you are already thinking that this all sounds too weird to be true, I haven't even started yet. We will see what I mean by all of this later on with plenty of examples. As a stopgap, we know that light, a form of electromagnetic radiation, is a form of wave, right? We also know it consists of discrete packets of energy called *photons*, which are a form of particle, i.e. it is both a particle and a wave. Confused yet? If not, then I haven't exposed just how confusing and counter-intuitive much of physics is, especially anything to do with quantum physics. You may be thinking, "ah, but photons have no mass, so they are not *really* particles", so this argument doesn't work. Nonetheless, particle or wave, all objects behave in odd ways at the nanometer scale whether we like it or not!

What does any of this have to do with nanotechnology? The message I would like to get across here is simply that over a hundred years ago, we thought we knew it all which clearly, we did not, and we should strive never to make that mistake again.

Now that we have a very brief overview of what we will be talking about, we can start to go a little further down the rabbit hole. We will take a chronological look at the story of nanotechnology starting with the history of the relevant areas of science, quantum mechanics, nanomaterials, computers, and finally nanomedicine. It will take a while to get a full appreciation for the breadth of nanotechnology, but it's worth the wait! I make no apology for taking my time and discussing related topics along the way as I see this all as part of the bigger picture which we do need to think about. So, dear reader — hold on and be patient!

Chapter 2

There's Plenty of Room
at the Bottom

If there is one phrase that has been beaten to death in nanotechnology, it is *there's plenty of room at the bottom*, which is an obscure way of saying "things can get an awful lot smaller than they are now and still work, so let's see just how small we can make them". In fact, I have seen and heard this phrase so many times that it has become a sort of earworm for me, like the lyrics of some songs. Listen enough times and you start to loathe it. Well, that's where I am with this particular phrase, despite the fact that it represents a pivotal moment in our scientific journey through the 20th century and has proven to be prophetic. In December 1959, the American physicist Richard Feynman gave a talk with this title to a meeting of the American Physical Society in Caltech. Before I divulge the contents of his lecture, we should first know who Richard Feynman was and why anybody listened to him give what would otherwise be considered as a fanciful science fiction talk.

Feynman was a Nobel laureate (in 1965) in physics who developed the theory of quantum electrodynamics (a rather complicated theory with the amusingly chosen acronym "QED") and which is used to describe the interaction of light with matter. As well as a theoretical physicist, he was also very well known as a bongo player, an artist, a lock-breaker, a decipherer of Egyptian hieroglyphics, someone who did things his way (i.e. a maverick with a healthy disregard for overzealous authority), and a brilliant communicator of complex ideas in physics. *The Feynman Lectures in Physics* are still essential reading for any aspiring physics undergraduate and were one of the first things I bought when I started at University back in 1988. He described himself however just as a theoretical physicist. He had honed his mathematical skills while working under the guidance of Oppenheimer on the Manhattan project — the development of the atom bomb, which had occupied many of the Allies' most brilliant physicists for a number of years in the early 1940s.

Although this endeavor helped to train many scientists, we must not forget that it ultimately led to a catastrophic loss of life. This has raised serious ethical concerns which will have reverberations for years to come. Feynman wrote a number of books for the general reader, ranging from amusing autobiographical works to treatises on complex ideas in physics, all of which sold very well. It was however in 1986 that he became a household name, at least in the USA, when he was a member of the Rogers commission that investigated the cause of the explosion of the space shuttle, the *Challenger*, on January 28th earlier that year. I remember having just turned 15 and watching it on the news later that day and thinking that maybe being an astronaut wasn't a great career choice after all. He demonstrated that the cause of the explosion was that the rubber seals on the fuel tanks had become brittle due to an unusually cold morning — i.e. the temperature dropped below the glass transition temperature of the rubber used to make the seals, so it went from being elastic and spongy, capable of forming a good seal, to brittle. The rubber then cracked and caused a fuel spillage under the stress of take-off. The leaked hot fuel gases burned through the tank and ignited the propellant, causing an explosion (*Source*: NASA document "Space Shuttle Era Facts") (Figure 1). The launch had been delayed a number of times by this point and NASA officials were keen to press ahead despite the temperature being lower than

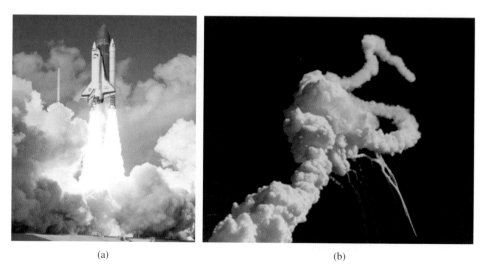

(a)   (b)

Figure 1.   (a) The Space Shuttle Challenger; (b) What happened when NASA tried to fool nature on the 28th of January 1986.

what was considered optimal and despite consequent warnings from NASA engineers that the launch should not go ahead. He ends his section of the report with the now-famous phrase: *For a successful technology, reality must take precedence over public relations, for Nature cannot be fooled*.

Given that there are over 2.5 million moving parts in a shuttle, any of which could cause a problem, it was decided to carry out a complete overhaul of the entire system. NASA spent the next two and a half years checking every component in the remaining shuttles, designed out the weaknesses, and built a new shuttle, the *Endeavour*, at a cost of $1.7 billion, resulting in safe shuttle missions for the next 25 years until the *Columbia* disaster in 2011.

By the time Feynman gave that talk at Caltech to a packed auditorium in 1959, he was known to have a sharp mind, and so people took heed when he talked about physics. The basic premise of his talk was to imagine we could make machines or devices as small as single atoms, and to think about what would happen if things were that small, without worrying too much about *how* they would be made. The integrated circuit had just been patented by Jack Kilby & Robert Noyce at Texas Instruments and would go into production the following year, in 1960. Transistors and other electronic components were still very large — on the order of millimeters across (Figure 2).

Nowadays transistors are around 15 nm across, which is only 60 atoms wide, so we are getting closer and closer to Feynman's suggestion. He outlined the various ways and means by which one might eventually be able to make things with sizes approaching atomic dimensions (which is what nanotechnology is all about), and the benefits and consequences of doing so. Now, we must remember that in those days — over 50 years ago — visualization of structures at these tiny dimensions was a dream as yet unrealized. Anything smaller than around 1000 nm

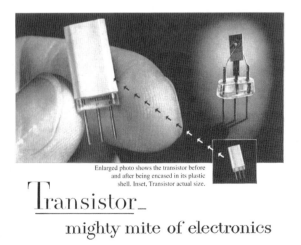

Figure 2.   An early ad for the semiconductor transistor, from the 1950s.

(1 micrometer, commonly just called 1 micron) is too small to see clearly using a normal optical microscope, so we need other forms of microscope to see them. The first of these alternatives to be developed, the scanning electron microscope, was not in common use yet. Despite the fact that mainframes were starting to be used, personal computers were not commonplace — the first, developed by Olivetti came out 5 years later (in 1964) and Intel wasn't formed until 4 years after that in 1968. Of course, the reason for wanting to make things so small was twofold:

(i) The smaller a thing is, the more of them you can pack into a certain space, which if we are thinking about transistors in a computer means that you can make a more powerful processor, or more memory without the computer having to get too big; this is how we can now carry around phones that are millions of times more powerful and faster than our computers of 20 years ago.
(ii) At atomic dimensions and just above, quantum effects start to become notice-able, which means you can make things that display novel behavior and *do different things than before*.

The second of these points in particular is what nanotechnology is all about. Feynman also had another, ulterior motive, which was to excite the next genera-tion of physics students. He did this by proposing a competition, subsequently known as "Feynman's challenge", where he offered $1,000 to the first person who could "take the information on the page of a book and put it in an area 1/25,000 smaller", and another $1,000 to the first person to create a working electric motor

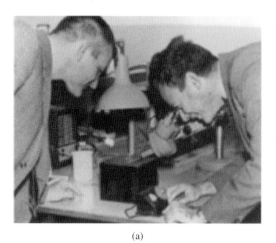
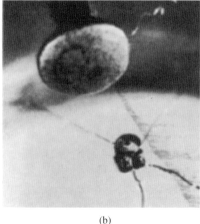

(a)   (b)

Figure 3.   Feynman's first challenge. (a) Feynman (the person on the right) looking at McLellan's tiny motor; (b) The motor as seen next to the head of a match.

that was no bigger than a 1/64-inch cube, or just under 0.4 mm on a side. It came as something of a surprise when the first of these prizes was claimed less than a year later by William McLellan, a local machinist, who made such a motor using standard tools (Figure 3). His trick was oodles of patience and a very steady hand. When Feynman saw it, he happily handed over the cash, although he did jokingly tell McLellan to not start writing small as he had just got married and bought a house, so had no intention of making good on the second challenge just yet, as can be seen in the letter that Feynman wrote to accompany the cheque he sent McLellan (Figure 4).

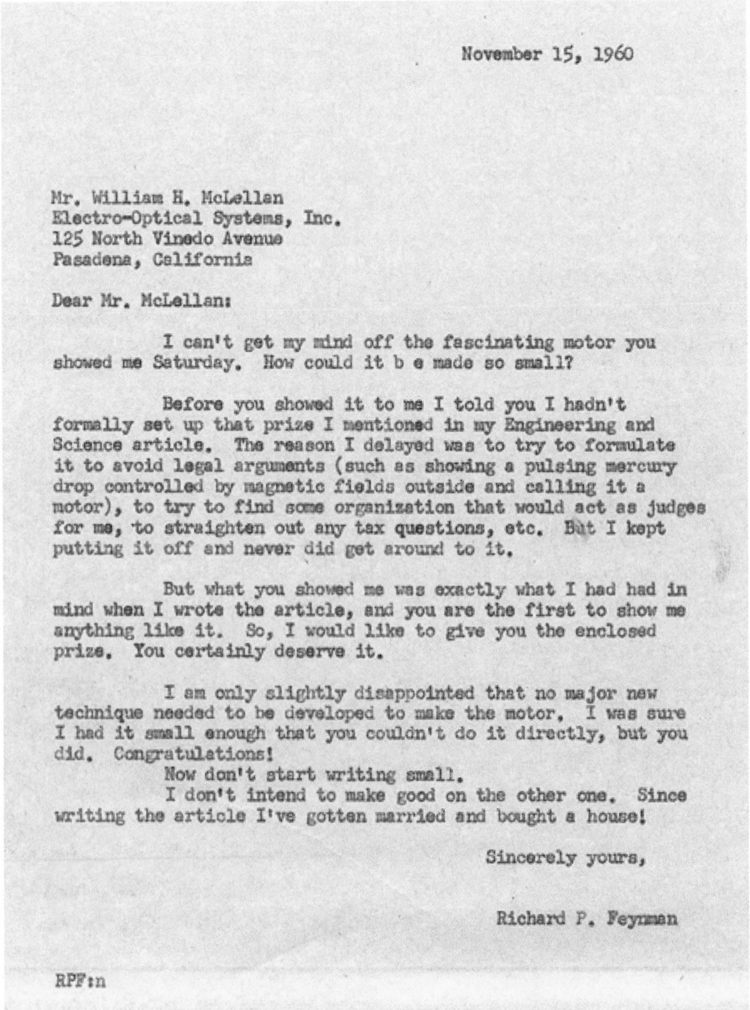

November 15, 1960

Mr. William H. McLellan
Electro-Optical Systems, Inc.
125 North Vinedo Avenue
Pasadena, California

Dear Mr. McLellan:

I can't get my mind off the fascinating motor you showed me Saturday. How could it b e made so small?

Before you showed it to me I told you I hadn't formally set up that prize I mentioned in my Engineering and Science article. The reason I delayed was to try to formulate it to avoid legal arguments (such as showing a pulsing mercury drop controlled by magnetic fields outside and calling it a motor), to try to find some organization that would act as judges for me, to straighten out any tax questions, etc. But I kept putting it off and never did get around to it.

But what you showed me was exactly what I had had in mind when I wrote the article, and you are the first to show me anything like it. So, I would like to give you the enclosed prize. You certainly deserve it.

I am only slightly disappointed that no major new technique needed to be developed to make the motor. I was sure I had it small enough that you couldn't do it directly, but you did. Congratulations!
Now don't start writing small.
I don't intend to make good on the other one. Since writing the article I've gotten married and bought a house!

Sincerely yours,

Richard P. Feynman

RPF:n

Figure 4.   Feynman's letter to McLellan congratulating him on cracking the first challenge.

Don't forget that $1,000 in 1959 is equivalent to over $8,000 in today's terms. The second challenge proved to be a much more difficult task that took another 25 years to be met, in 1984 by Tom Newman, a graduate student at Stanford. By using the finely focused electron beam of an electron microscope to basically etch tiny holes in a polymer film, he reproduced the first page (with the total size of around 15 microns, each letter being a few hundred nanometers across) of Charles Dickens' *A Tale of Two Cities* on a pinhead and was able to claim his prize (Figure 5(a)). The biggest problem he faced was the fact that it was inordinately difficult to locate the page he'd written as it was so small! By the time Newman had done this, the 66-year old Feynman was very happy to hand over the $1,000 as Newman had to apply a new technology in order to meet the challenge.

The technique that Newman applied, *electron-beam lithography*, had been invented in 1960, but was not in common usage until the late 70s. It is now a finely tuned tool, capable of creating structures as small as a few nanometers across — central to the continued success of the semiconductor industry. The next major breakthrough along these lines came in 1989 when researchers at IBM in Almaden wrote their company name using single atoms, with the aid of the recently invented (by IBM researchers in Zurich) Scanning Tunneling Microscope, or STM. In fact, it was this very image that inspired me to become a nanotechnologist — I was an undergraduate at that time, wondering about what area of physics to try to get a job in and was seriously considering particle physics until I saw that image (Figure 5(b)).

Unfortunately, Feynman did not live to see his dream of building things atom-by-atom come true, as he had succumbed to cancer on February 15th the year before at the tender age of 69, but he knew that this was just around the corner. He did not go out without a sense of humor, as his last recorded words were: *I'd hate to die twice, it's so boring.* Nonetheless, many of his predictions about

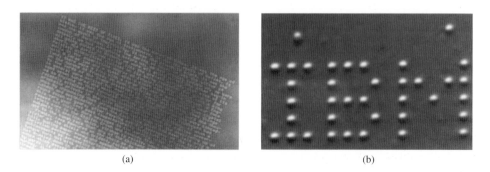

(a)   (b)

Figure 5.   Feynman's second challenge. (a) Opening passage from "A Tale of Two Cities" written at the nanometer scale — achieved in 1984 (*Image Courtesy*: Caltech Library); (b) The ultimate in precision manufacturing — "IBM" written using single atoms (each bright dot is one atom), from 1989 (*Image Courtesy*: IBM).

Nanotechnology have come true, so his insightful talk back in 1959 has served as a valuable roadmap, at least in hindsight.

Feynman's talk is now the most commonly cited in the history of Nanotechnology, even though the field really kicked off over 20 years *after* he gave it, and it only became widely known by that generation of scientists after the publication of Eric Drexler's book on molecular machines (see Chapter 2) in 1986 wherein Drexler took Feynman's ideas quite a bit further and into the public domain.

Tools of the Trade, and 10 Magic Years

Nanotechnology has been on the cards for a very long time, but there is one event that single-handedly led to its rise to prominence, which by the way was *not* Feynman's talk. This was the aforementioned invention of the STM in 1981, by IBM scientists Heinrich Rohrer and Gerd Binnig, with the help of the most ingenious research scientist I have ever encountered, Christoph Gerber, who has been responsible for many of the key developments in nanotechnology over the last 40 years, many of which we will touch on in this book. Gerber was duly honoured for his seminal contributions with the award of the Kavli Prize for Nanoscience in 2016. With this microscope, it became possible to image things at the scale of single atoms, and also to *move* those atoms around and build new structures one atom at a time. Five years later (in 1986 — the same year that Binnig & Rohrer were awarded the Nobel Prize in Physics), Binnig, Gerber, and Calvin Quate from Stanford went on to invent the Atomic Force Microscope (AFM), which, together with the STM, revolutionized materials science, physics, and chemistry by allowing us to truly see the properties of matter at the atomic scale and opened up a whole wonderland of opportunities in materials research. We will look at these microscopes and the magic they can perform later, but for now let's be content with the knowledge that there are a family of tools that can be used to explore the properties of materials down to the atomic scale that have led to the blossoming of Nanoscience and Nanotechnology. I was 15 when the AFM was invented, and I saw a picture of some surface taken with one in the *Irish Times*. This made a lasting impression on me and helped me to see that science was interesting and worth pursuing, especially if you are naturally curious.

At around the same time, two new, nanometer-sized forms (allotropes) of carbon were discovered, adding to the frisson of excitement surrounding nano-things and ultimately avalanched into what we know today as nanotechnology. In 1985, Buckminsterfullerene (or more simply, carbon-60, or C_{60}) was discovered by Richard Smalley's group at Rice University. Even that is a story that beggars belief. Smalley's group was studying a problem in astrochemistry — one can determine the composition of a star by looking at the spectrum of light from it.

Different materials give off different colors of light when heated, so by looking at the colors (as well as those that are missing), it is possible to figure out what materials are present. They were unable to explain where a particular set of colors was coming from and, on the basis of some calculations, proposed that they must be coming from stable carbon particles containing 60 atoms, now known as C_{60}. Its existence had been proposed but not proved somewhat earlier by Osawa from Japan in 1970 who was studying molecules. The C_{60} molecule is a hollow sphere, with the carbon atoms arranged in hexagons and pentagons similar to that found on a stitched leather football and has a diameter of 0.7 nm. This molecule aroused a lot of interest as it had some interesting electrical properties, was superconducting at low temperatures, and was used to encapsulate all sorts of atoms, but eventually interest waned when applications failed to materialize — a common issue as we will see. However, the approach developed to explore its properties has turned out to be invaluable when applied to other materials since then. In 1976, hollow carbon fibers with nanometer diameters (i.e. nanotubes) were discovered by Morinobu Endo, who patented their production, but this blended into obscurity until 1991 when Sumio Iijima from NEC in Japan published a paper on his discovery of Carbon Nanotubes, which he described as hollow cylinders of carbon with a wide range of diameters starting from as small as around 0.3 nm (Figure 6).

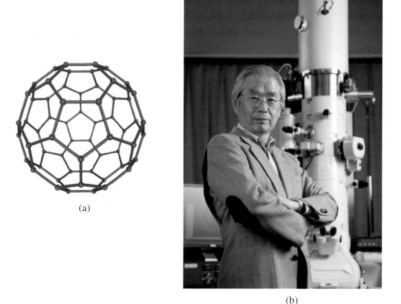

(a)

(b)

Figure 6.   (a) A C_{60} molecule, otherwise known as a buckyball; (b) Sumio Iijima, whose discovery of carbon nanotubes inspired many people to be interested in Nanotechnology.

We will look at these in later chapters when we will also look at carbon fiber, but carbon nanotubes first gained interest as they have some impressive electrical and mechanical properties, and a vast amount of research went into trying to make a plethora of electronic devices using them. Unfortunately, this also waned as reproducibility was an issue, and potential toxicity concerns put paid to most research in the early 2000s. However, carbon nanotubes have rather excellent mechanical properties, which have turned out to be very useful in the field of composites, and they are widely used today to make strong and light materials.

Structures similar to carbon nanotubes, although rather large at around 50 nm in diameter, had been reported by Russian scientists back in 1952, but these were considered less interesting than the approximately 1 nm diameter nanotubes shown by Iijima and so were largely forgotten.

Rediscovering the Past

This general theme of things being discovered, or rather re-discovered, has been going on for some time to say the least. For instance, it is now known that, somewhat unwittingly, carbon nanotubes were first made in at least 300 BC, in the process producing so-called Wootz steel ingots in India. This steel was used to make Damascus steel — the material of choice for the best quality, sharpest, and hardest blades in antiquity. A high proportion of carbon-based precursor materials (i.e. wood) were used in the formation of Wootz steel, and the particular heating (annealing) and forming conditions used to make this steel led to the formation of carbon nanotubes, to which the advantageous properties of the material, which we could argue was possibly the first composite, have been attributed. The presence of carbon nanotubes in such a blade was only reported quite recently in 2006, where they were revealed by powerful microscopes on a 17th century blade made by the Persian master blacksmith Assad Ullah, using the old techniques from India.

A little later, the Romans were using silver nanoparticles to create interesting color patterns in precious glass objects — early stained glass. The best-known example of this is the Lycurgus cup, a 4th century AD drinking vessel that appears to have different colors depending on the angle of illumination — when light is reflected from it, it looks green, whereas when light is transmitted through it, it looks red (Figure 7). This is all down to the way that silver nanoparticles absorb light, and in the absence of any form of microscopy, the Romans did not know what it was that gave rise to this effect, except that it was aesthetically very pleasing. Another similar example is Maya blue, a pigment from 8th century Mexico that uses clay with nanometer-sized pores, into which indigo is mixed, forming a brightly colored paste. The nanoporous clay binds the dye and stabilizes it, helping

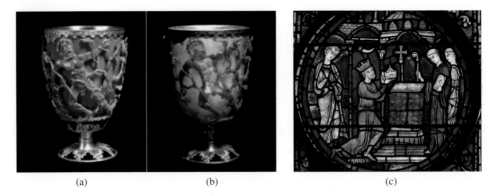

(a)   (b)   (c)

Figure 7.   (a)–(b) The Lycurgus cup — a 4th century Roman example of the use of nanoparticles to control color. The red and green are the apparent colors when light is reflected from or transmitted through the glass, respectively. The color comes from gold and silver nanoparticles in the glass; (c) Stained-glass window from Chartres cathedral in France, circa 13th century. The colors are all due to gold nanoparticles incorporated in the glass.

it maintain its color for over 1,200 years, which would not happen otherwise. The use of nanotechnology continues to this day in stained-glass windows, with the earliest known examples containing silver nanoparticles (via a process imaginatively known as silver staining) dating from the 14th century (Figure 7). I would argue that while these historical examples demonstrate that nanotechnology has been in use for millennia, the people involved were not nanotechnologists with all due respect, as they had absolutely no idea what they were doing, Nanotechnology is all about doing this kind of thing *on purpose*!

The Nano Approach to Creating Solutions

In the most general sense possible, when it comes to manufacturing things, there are broadly two different approaches that are taken: *top-down* and *bottom-up*, sometimes called *subtractive* and *additive*, respectively (Figure 8). Top-down is where you start with something big and carve it into the final object you want, like molding or sculpture. Bottom-up is when you build something bit-by-bit, for example, when a complex machine is made by putting individual components or building blocks together. Conventional manufacturing typically uses both — for example, when cars are made, the chassis, the frame, and the individual components are often made in a top-down fashion, and then the lot is assembled, which is a sort of bottom-up process. From a nanotechnology perspective, when we talk about a bottom-up approach instead of making something bit by bit, we mean *atom-by-atom* or *molecule-by-molecule*. We will see some examples of this later, and how they can be and are applied today. As a teaser, DNA and

(a)   (b)

Figure 8.   (a) Sculpture is an example of top-down (subtractive) fabrication — starting with a piece of material, and removing the unwanted bits; (b) A large-scale example of bottom-up (additive) fabrication — building something using building blocks that get stuck together.

proteins are naturally made in a bottom-up process in the form of a chain of atoms that then folds into the desired shape (hopefully, as misfolded proteins are associated with many degenerative diseases). Integrated circuits are made in a top-down process where silicon is processed and patterned until it has reached the desired shape, size, and overall composition to be able to control the flow of electric current in just the way we want. There are several hundred steps involved and the entire process is enormously complicated, so anything that can simplify this would be welcome.

We will see later that bottom-up fabrication is a key technique used in nanotechnology, and the semiconductor industry has flirted with it in an attempt to simplify circuit manufacture. Before delving any further into this, we need to get to grips with our past and have a rapid refresher course in the history of physics.

A Brief History of Physics

Feynman's contribution to nanotechnology was really to start the ball rolling on getting scientists to think more outside the box. This was the root of his special talent for physics — he had an atypical way of thinking, so saw solutions to problems with a clarity that many others did not. His natural curiosity led him to question why things are the way they are, and his deep knowledge and understanding of physics allowed him to forge new paths at the heart of quantum mechanics. However, he was, as is always the case, standing on the shoulders of the scientific

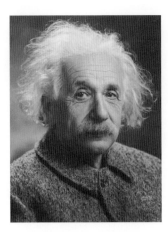

Figure 9.   Albert Einstein — who re-wrote the rule book of physics in the early 20th century.

giants who had gone before. In 1904, just 55 years before Feynman delivered that talk, Einstein (with others) had turned physics on its head. Einstein was another character who was not afraid to challenge the conventional "wisdom" and through his insights into the nature of almost all of physics, essentially threw away the old physics of Newton (Figure 9).

In the course of one year (the "Annus Mirablis" meaning *remarkable year*), he set the foundations for modern physics as laid out in four papers of his that year. In them, he introduced:

(1)   special relativity,
(2)   quantum mechanics (he used it to explain the photoelectric effect: whereby when light is shone on a metal surface, electrons are emitted from that surface — this was the work for which he won the Nobel Prize in 1921),
(3)   the mass–energy equivalence, i.e. $E = mc^2$, and
(4)   he explained the molecular processes behind Brownian motion (the random motion of particles in a fluid due to collisions with the fluid molecules).

The vast majority of modern physics stems from one of these papers of Einstein or those that followed. Feynman's great triumph was to combine quantum mechanics and electromagnetic theory for a particular set of circumstances, which is what QED is all about.

You might imagine that understanding how light interacts with matter should not be that difficult to understand. However, it is fiendishly complicated as quantum mechanics which describes how matter behaves and electromagnetic theory which describes how light behaves had been considered to be mutually exclusive until then.

The past 100 years has seen a total renaissance in physics, and in fact all of Science, and new developments are constantly being made, from the discovery of (1) the Higgs Boson which is the master particle responsible for other particles having mass — predicted in 1964 by Peter Higgs, and discovered in 2012, to (2) gravitational waves which are ripples through spacetime caused by the massive inflation of the universe shortly after the Big Bang — they were predicted by Einstein in 1916 and discovered in 2014/15. When Michelson made his remark about physics basically being known, he had a point as very little had happened apart from more and more precise measurements during the previous 250 years. Perhaps there is a lesson in that — past performance is no guarantee of future performance. We all learned that in a very painful way in 2008 when global markets crashed, against many predictions to the contrary. The previous reset in physics came when Newton devised his three laws and used them to describe gravity. At this point, a brief retelling of that story will help us to better understand the longer-term cycles in science and therefore be somewhat more circumspect in our attempts at predicting the future.

Our ancient ancestors realized the earth went around the sun in an approximately circular path that takes one year. We know this from Egyptian, Greek, and Chinese writings/pictograms. Thousands of years ago before very large cities were formed and before there was much by way of light or air pollution, the heavens would have been far more visible at night for many people than they are for most of us today. The positions of the stars and other heavenly bodies were considered to be a significant factor affecting people's fortune and daily lives, so a great deal of time and effort was expended on mapping the skies and the motion of everything visible in them (Figure 10). Today we call this *astrology* and I find it highly ironic that for the vast majority of people, the perceived usefulness of our detailed knowledge of the stars and planets and how they move is that "better" horoscopes can be written. It is deeply rooted in both eastern and western culture that our star sign plays a very large role in determining our fate. Where does this come from? It is difficult to understand how totally rational individuals can place any stock in this idea apart from the human desire to feel that we have a purpose. Astronomy and astrology are often confused for each other when they really couldn't be much further apart as far as mindset goes. Somehow with the advent of western science thousands of years on, this original observational information from astronomy was lost in the middle ages and all had to be rediscovered in recent centuries (Figure 11).

This tale starts with Copernicus who in the early 16th century used his telescope to measure the motion of the planets. From these measurements, he

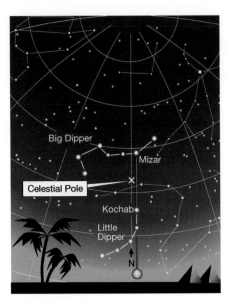

Figure 10.   The ancient Egyptians were highly adept astronomers and used positions of stars to align their buildings.

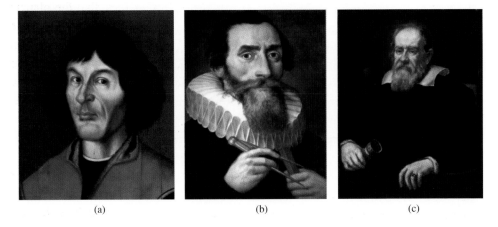

(a)   (b)   (c)

Figure 11.   The western scientists responsible for doing the legwork that allowed Newton to develop his law of gravity. (a) Copernicus, circa 1580; (b) Kepler, circa 1610; (c) Galileo in 1636.

concluded that the sun was at the center of the solar system rather than the earth, and that the planets followed circular paths around the sun. Then, in the early 17th century, Brahe and Kepler used more powerful telescopes and took more detailed measurements and concluded that the planets went around the sun following slightly elliptical rather than circular orbits with the sun at one of the foci, and Kepler deduced the laws describing the effects of gravity.

At around the same time, Galileo fine-tuned these measurements as he built even better telescopes with higher magnification, enabling more precise measurements of the positions of the stars and planets. He was vocal in his findings and as they were at odds with the Catholic church's teaching of the time that the earth was at the center of the entire universe, he was found guilty of heresy and committed to house arrest for the rest of his life. He escaped the death penalty by renouncing his findings. His measurements were in fact so accurate that he noticed Jupiter was not following a perfectly elliptical path, but that it was periodically swaying a little off course. He suggested that there had to be something periodically pulling it off course, and that that something would probably be one or more moons orbiting around the planet. Seems like a perfectly sensible thing to say now, but at the time, this sparked quite a bit of controversy. He observed the moons of Jupiter shortly thereafter, and indeed it was found that the center of mass of Jupiter with its moons was following an ellipse around the sun, but that as the moons orbited the planet, it wobbled a bit. Similar arguments continue today when it was noticed that on a much larger scale, galaxies move in a way that is slightly inconsistent with their known mass, which led to the suggestion that they have more mass that we can't see for some reason — hence the name *dark matter* came to be coined.

Within a few decades of this, Newton was developing as a scientist and he was the first to figure out, using logical arguments and mathematics, that the planets would indeed follow elliptical paths around the sun. He also showed that the time taken for a planet to orbit the sun would be related to how far away it was from the sun and that the force holding the planets in those orbits was proportional to the mass of the planet and the sun, and inversely proportional to the square of the distance between them. In this way, he formulated the law of gravity for the first time and developed a mathematical framework, including calculus, that has been used to this day and is standard fare for anyone studying physics or applied maths.

The question as to what exactly gravity was and where it came from was a separate matter entirely, and to be honest, we don't really know — we have better descriptions of what it does, but not what it actually *is*. The ancients had all sorts of theories as to how the planets moved around the sun, my favorite being the idea that there were angels pushing them around. What Newton showed was that the force on the planets was not in their direction of motion but was in fact *towards the sun*. In other words, gravity is what is known as a *central force*. It is the same force that is experienced by a stone as it is swung around in a circle at the end of a piece of string or by a car as it drives around a corner. If the string breaks or the road is icy, the stone or the car will stop going around in a circle, and head off in a straight line — a force directed toward the center (called the *centripetal* force) is needed to keep them moving in a circle. In all cases, there must be an opposing force, to stop these things from moving outward. In the case

(a)   (b)

Figure 12.   (a) Isaac Newton who formulated the rules of mechanics in the late 17th century; (b) Centripetal force at work — these racing cars experience a force directed outwards, as indicated by the arrows. The opposing force (i.e. what stops them from moving outwards) is provided by friction between the wheels and the ground.

of the stone, it is tension in the string, and in the case of the car, it is the friction force between the wheels and the ground. Given how fast racing cars are going (well over 100 mph when cornering), the friction forces are considerable, which is why the wheels need to be changed so frequently as they get worn out. Newton devised his famous three laws of mechanics related to inertia, force, and action and reaction, and for the next few centuries, this is what physics was all about (Figure 12). That is, until Einstein came along and rewrote the rules in the form of general relativity.

Einstein's starting point was that he wanted to understand *action at a distance*. The laws of physics at that point suggested that if something changed in the universe, it's effect would be felt everywhere *instantly*. He saw that this would mean information was transmitted infinitely quickly, which he doubted, and upon exploring a way around this, he discovered that information does in fact travel at a finite speed and is limited to the speed of light. This meant that we had to modify our understanding of how both space and time work and led naturally to the concept that gravity is a *distortion* of both space and time ("spacetime") which is caused by mass, and the more massive (heavy) an object is, the more it distorts spacetime (Figure 13). He found that this was important for describing the behavior of large and heavy things such as stars, and things that are traveling very fast, close to the speed of light. This way of viewing gravity is not for the faint-hearted, but it does provide predictions that have stood the test of the last hundred years. I should put in a disclaimer here, as nature is unfortunately not

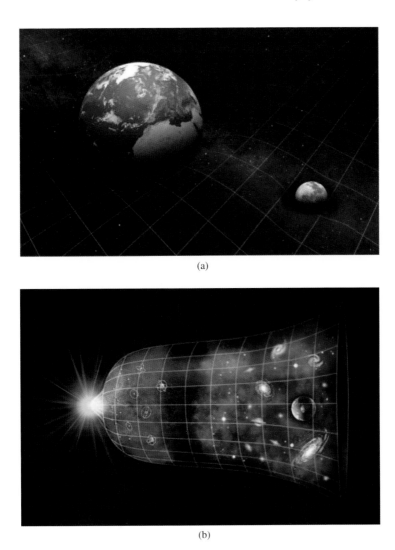

(a)

(b)

Figure 13.   (a) Spacetime, here illustrated as a mesh, is distorted by objects with mass. This warping of spacetime is what we experience as gravity; (b) Artist's rendition of the big bang. Time is increasing from the left.

without a warped sense of humor. In the quantum description of things, we *do* get instantaneous effect for any cause, i.e. the speed of light is not a limit. As an example, we know that it takes a little over 8 minutes for light from the sun to reach earth, as it travels at a finite speed. However, if there were an object on the sun that was related to an object on earth via a quantum interaction, then a change in one would produce an *instantaneous* change in the other. As we will see later, this is not science fiction, and such quantum effects happen all the time.

Einstein realized this as well and called it *spooky action at a distance*, and a multitude of experiments over the past few years have shown it to be true. This curious and deeply unsettling topic will pop up again.

The mathematics behind relativity is rather involved, but the fact is that for most things we observe in our daily lives, Newton's laws are good enough. This does not include the nanoscale, but Einstein comes again to the rescue. As mentioned earlier, he was instrumental in devising quantum mechanics, which is the physics of the very small, and highlights that nature is far more complicated and terribly strange than we can truly comprehend. When we look around us in our daily lives, things do not behave in a quantum way, so whenever we do see quantum mechanics in action, it is like looking through a distorted window — things just don't make sense!

The bottom line with quantum mechanics is that things are not entirely *deterministic*, i.e. we can't predict what will happen and when, all we can predict is the *likelihood* (physicists prefer to use the word *probability*) of things happening. This deeply troubled Einstein who did not like it, as his religious conviction was that "God does not play dice" (Figure 14). He did however realize when he devised relativity that it was not the complete picture and he could not reconcile gravity with quantum mechanics. He also recognized that even space itself was an entity that needed to follow the rules of quantum mechanics, albeit at very small lengthscales (10^{-33} m) and could find no way to add this to either theory. This is a feat that has still not been achieved apart from a very few examples. We will explore the intricacies of quantum mechanics in the next chapter, as this will help us to understand why materials behave so differently when they are nanometer sized. The phenomenon behind all of this is the fact that particles and waves are

Figure 14.   Quantum mechanics, due to its probabilistic rather than deterministic nature has been likened to "God playing dice" by Einstein, who was deeply troubled by the implications.

both aspects of the same thing, and this is particularly relevant at very small length-scales — nanometers.

As a brief aside, for a similar reason, when he devised the equations of general relativity, he saw that they predicted the Universe was either expanding or contracting, but is certainly not static. He did not like this either (again, on religious grounds), so he added a fudge factor that he called the cosmological constant that would correct for this. In later life, he recounted that he considered this to be his greatest mistake. All the evidence is of course that the universe is expanding and has been since the Big Bang, and that it will continue to do so. Therein lies the beauty of scientific theories based on mathematical principles. If the initial insight that led to the development of a theory is sound, then we need to put aside any preconceptions when looking at the predictions of that theory. The mathematics does not lie, but we do when we start trying to interpret or alter things to fit with our own limited understanding.

In order to understand how nanotechnology works, while we do not need to know all that much about quantum mechanics, we do need to be aware that very small things behave differently to bigger things and that the particular way we need to look at the world around us depends on the context and the size of what we are looking at. In the next chapter, we will take a look at the developments that led to quantum mechanics, and the far-reaching consequences this has had in so many aspects of our daily lives.

Chapter 3

The Science Bit

It is often stated that of all the theories proposed this century, the silliest is quantum theory. In fact, some say that the only thing that quantum theory has going for it is that it is unquestionably correct
— Michio Kaku, co-creator of string theory

I think I can safely say that nobody understands quantum mechanics
— Richard Feynman, Physics Nobel laureate for quantum electrodynamics

We have seen that nanotechnology is all about the application of science and engineering at the nanometer (1 (UK) billionth of a meter) scale. This is an *atomistic* scale, i.e. comparable to the size of atoms, and it turns out that the properties of matter at this scale are best described through the use of quantum mechanics, so we will spend this chapter unravelling that in order to then *understand* rather than just *know* what nanotechnology is all about and also what it can and cannot do (Figure 1). Quantum mechanics and its effects are all around us and we experience them in our daily lives without even realizing. We have had over 100 years of exposure to the weird and confusing way quantum mechanics makes us look at the world and are still just unable to truly comprehend what it means, as alluded to by the quotes above. They may seem flippant, but they sum up the current state of our knowledge perfectly.

By its very nature, science is always in a state of flux as our knowledge and understanding of the world around us is constantly evolving, as are our theories or models describing things.

The purpose of scientific theories is to produce a set of testable hypotheses. As soon as experiment shows a deviation from what a theory explains, if we are satisfied that the experiment was carried out properly and carefully and we have taken into account all the things that could give rise to an anomalous result, then

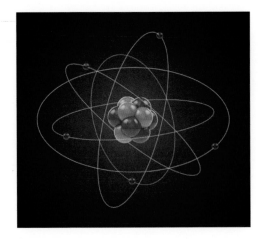

Figure 1.   Quantum mechanics is used to describe the properties of atoms, which are the building blocks from which stuff is made.

we must discount that theory and start again. A theory is only useful if it agrees with experiment, no matter how mathematically perfect it may be. There is a bit of a thing going on there in many cases — experimentalists often think that theory/ modeling is not very useful as it is usually wrong or is so grossly oversimplified that it could not describe something in reality. At the same time, theoreticians often feel that experimentalists get their hands dirty and if they get an odd result, they just assume the experiment was not done properly! There are times however when they work in tandem and support each other and together genuinely improve our understanding of how things work. I have been fortunate enough to be involved in several such collaborations that have borne fruit.

The better a theory is able to explain and predict phenomena, the more widely it becomes accepted. Sooner or later though, all theories turn out to be mere approximations to more complex, more fundamental theories, hence the ill-fated (so far at least) quest to find a *master theory of everything*, or as Einstein called it a "grand unified theory". There is no particular reason to expect that such a theory exists, but it would be very convenient if it did! In the time before rigorous logic was applied to understanding the world around us, our cultures were riddled with what we now consider to be bonkers or at least decidedly odd ideas. Chief among these were the notion that the earth was flat, the earth was supported by an elephant on top of a tortoise (yes…), the earth was pushed around the sun by angels, and basically anything that was outside our understanding was either ascribed to some deity or another or considered to be magic.

For better or worse, we live in a world where the general consensus is that most things can be explained by science, so there is little room to resort to explaining things away by magic, even if we do not know the answers to everything yet. We have arrived at this point due to the incontrovertible success of science at describing the world around us. There are still unanswered and possibly *unanswerable* questions, such as where did we really come from, what spark created life, are we alone in the universe, what happened before the big bang, if you were to go to the edge of the known universe, what would you find there, and what is quantum mechanics all about? Obviously, it would be folly for me to even *consider* any but the last question. Science has had a number of successes over the last few centuries and I will focus on the basic underlying principles. These have led to demonstrable changes in the way we do things or in what we can do, and which formed the pillars upon which nanotechnology has been built. The first of these was Newton's work on mechanics. Schoolchildren across the world have at least heard of Newton's three laws of mechanics. My own children practically groan whenever I mention them as I do so very often, apparently. Well, that's because they are so useful! Whenever my son plays football, or my daughter hits a tennis ball, we think about the forces that are acting, the effect that they have, and how, with a bit of maths, we can predict what's going to happen. When they are older, I will show them how to write down the equations describing those forces so that they can see exactly how to calculate the exact trajectory of said ball. If they ask nicely, that is. Of course, to solve those equations requires the use of calculus, which Newton also developed for that very reason.

There is a tendency for many people to glaze over whenever equations are even mentioned, and when Steven Hawking was writing *A Brief History of Time*, his editor told him that every equation he put in would halve the book sales. He decided it was worth taking the risk, as he left one equation in — that $E = mc^2$. Of course, it is possible to explain a concept in physics without using any equations, but to do so is very limiting. The power of this relationship of course is that it tells us it takes an enormous amount of energy to create mass, and conversely that if we can convert just a tiny fraction of mass into energy, we could generate rather a lot of energy. This is what happens in nuclear processes — when two atoms are fused together as in nuclear fusion, the mass of the new nucleus is slightly less than the mass of the two nuclei added together, and this difference in mass is released as energy (kinetic energy of motion of the new nucleus), which is why we want to develop nuclear fusion as a power source (Figure 2). Something similar happens with nuclear fission as happens in power stations and atomic bombs. Einstein could have just said that the energy associated with any object is proportional to its mass. To be more exact, it is equal to the mass times the square of the speed of light. I don't know about you, but I find it much easier to just write the formula! Once we have the appropriate mathematical language to describe a system, we can manipulate the equations to explore how the system behaves and

 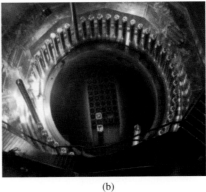

(a) (b)

Figure 2. (a) Einstein's mass/energy equivalence at use; (b) The core of a nuclear reactor — a somewhat more palatable application of $E = mc^2$.

then try to interpret what the maths is telling us. Of course, we must remember that the purpose of a formula is to capture an insight, which would have come first, and we then need that insight in order to apply the formula. The beauty of Newton's laws is that they are incredibly simple and make sense. They can be written in words as follows:

Newton's Laws

- **Law 1** — Any object will change its direction or speed if a force acts on it. This general property is known as *inertia.*
- **Law 2** — If we apply a force to an object, it will accelerate by an amount proportional to that force.
- **Law 3** — Every action has an equal and opposite reaction.

These laws are so pervasive and so commonly quoted that they seem like common sense. The power of them however, only comes to the fore when they are expressed mathematically, and we can then use them to not only describe the forces acting on any system, but to *predict the effect of those forces on how the system behaves*. For example, Kepler used observations combined with Newton's laws to derive the mathematical formulas describing the motion of the planets, and a whole new set of interesting features emerged from the maths that allow us to predict the orbit of any object based on a few simple observations.

So, if Newton's laws are so great, why do we need anything else? The fact is we didn't for over 250 years, as until then, experimental observations of the motion of objects could be explained and described perfectly using these laws. No doubt ignorance was bliss. During this intervening period, the branch of physics

Figure 3. Lightning is an electromagnetic phenomenon, as described by Maxwell's equations of electromagnetism, and the entire telecommunications and electrical industry uses phenomena explained by these equations.

associated with Newton's laws which was called *mechanics* flourished, and found many applications, especially in the design of buildings, structures, and machines. In the early 19th century, physicists then turned to the next problem — the behavior of electric and magnetic fields, and using Newton's calculus, they were able to describe these fields and their relationships to each other. This culminated in Maxwell's equations, and the branch of physics we call *electromagnetism,* that has given us telecommunications, radio, TV, microwaves, lasers, and basically everything electrical in common use today (Figure 3).

These equations were found to make predictions that also agreed perfectly with experiment, at least at the time. What changed then toward the end of the 19th century was our ability to make more and more precise measurements, until it got to a point when measurements were *so* precise that differences between experiments and theory started to be noticed. This was true both in the fields of mechanics and electromagnetism. Now, I should make it clear at this point that the reason for the success of Newton's laws was that they are very good approximations at the distances and speeds we encounter in our everyday lives, and they only really start to become ineffective at speeds approaching the speed of light, which is 300,000 km/s, and at distances typically greater than hundreds of thousands of kilometers. It was also noticed in measurements of very big things with large gravitational fields (galaxies), and very small things such as subatomic particles, that these familiar laws started to break down. By break down, I mean that the results of experimental measurements did not quite agree with what theory predicted. The motion of objects under these conditions simply does not follow that predicted using Newton's laws. This is not to be mistaken for chaos theory, where a tiny change in initial conditions produces a profound change in outcome, otherwise known as the butterfly effect. This name was coined by Edward Lorenz and encapsulates the results of chaos theory — that a butterfly flapping its wings somewhere could cause there to be a thunderstorm somewhere else weeks later, such is the delicateness of chaotic systems. This is the main reason why we cannot predict weather accurately beyond 4–5 days, although in practice the weather is not influenced by butterflies! As we don't tend to encounter galaxies, subatomic particles or objects traveling at light speed on a daily basis, Newton's laws are therefore perfectly adequate for most applications. They apply when we want to consider the forces acting on a football, a tennis ball, or a snooker ball; they can be used to describe the forces acting between a wheel and the road, on an airplane wing, and many other things. At around the time that Michelson made his famous remark about the future of physics ("there is nothing new to be discovered in physics now. All that remains is more and more precise measurement"), things were just starting to not look all that rosy in the world of physics. Measurements were showing that the orbits of the planets in our solar system had slightly different shapes than expected, and that this could not be explained by Newton's laws; that the newly discovered electron had some rather unusual properties — particularly when in a magnetic field; that nobody could explain why some materials were good conductors of electricity and heat and others were not, and nobody could adequately explain why materials have the color they do.

These are just the tip of the iceberg, but they are big questions that were not satisfactorily answered at that point in time. I would argue that the veil was drawn apart by Einstein in 1904 when he published those four papers that laid the foundations necessary to answer these questions. Firstly, through his ideas of relativity, he showed that time and space are related — what this really means is that if you measure the passage of time or the size of something, it depends on where you are (i.e. if you are near a strong source of gravity such as a star or a black hole) and on how fast you are going, which is completely counter-intuitive, although it was made to look convincingly simple in the movie *Interstellar*. As you get faster and start approaching the speed of light, any measurements you make of time or distance start to diverge significantly from those of someone who is stationary or moving slower than you. Einstein's theory was nothing short of revolutionary and met with considerable resistance initially, as it seemed to just not make sense. The maths was flawless but many physicists lacked the insight to believe it's foundations were correct. The best example of this is that you know if you are driving along the road at 50 mph, and you meet a car traveling in the opposite direction, also at 50 mph, then your *relative* speed is

100 mph, right? Right. What Einstein showed is that it is impossible for anything to travel faster than the speed of light, so if you are traveling at say, 3/4 of the speed of light and you encounter someone traveling at the same speed in the opposite direction, instead of your relative speed being 1.5 times the speed of light, it is just *below* the speed of light. How can this be? He showed all of this mathematically, and as maths is built on logic, everybody who read his papers could find no fault in them, and concluded that they must be correct, even though they didn't *feel* right. Now, that is oversimplifying things just a bit, as the most important ingredient for a theory to be accepted is that *it must agree with experiment*. To this day, the agreement between theory and experiment on this topic is pretty much perfect.

The icing on the cake came when Einstein showed in 1915, when he developed the theory of General Relativity, that gravity can be described as a bending of space by objects with mass, which has the direct prediction that light will change direction if it passes by any massive object. This had previously been predicted by Cavendish and Soldner independently in 1784 and 1801, respectively, on the basis of Newton's laws, but their predictions were out by a factor of 2. In 1919, Eddington led a mission to the island of Principe off the coast of Africa to test Einstein's idea — he looked at how the apparent position of a star changed when the light coming from it passed near to our sun (Figure 4). As the sun is so bright, this is only possible to observe during a solar eclipse, hence the exotic location for the experiment at that time. Needless to say, his experiments

Figure 4.   One of Eddington's photographs of the solar eclipse of 1919, from which the apparent change in the position of stars was measured and found to agree with Einstein's predictions, thus validating the general theory of relativity.

verified that Einstein's prediction was correct, with high accuracy. Relativity also explained a wobble in Mercury's orbit that could not be accounted for using Newton's laws and had been observed in 1859, and it also explained how light shifts its wavelength toward the red end of the spectrum when passing through an area of high gravitational field. This was not fully verified experimentally until 1959.

Shortly before Einstein came up with this idea, the seeds of quantum theory were being sown by Max Planck. He was trying to understand how objects absorb and emit heat. There is an idealization of this in physics, known as a *black body* — a fictitious object which absorbs all the radiation incident on it, and which therefore appears black. When such an object is heated up, it emits so-called *black-body* radiation. In other words, if you heat an object up, it starts to emit light — you know if you stick a poker in the fire, it starts glowing red, and if the fire gets hotter then the poker will turn yellow and eventually white. The shift in color as temperature changes is just showing us that the distribution of emitted light is tending toward shorter wavelengths for higher temperatures. All objects lose heat by emitting light, but it is only visible to the human eye for high tempera-tures (when the emitted light is within the visible spectrum) — thermal imaging cameras can detect much cooler objects. The reason why any of this is important is that there was a glaring hole in physics which was not able to explain *why* the color of light emitted from an object depended on its temperature. In fact, the theo-ries in place at the end of the 19th century predicted that black bodies would emit light at all wavelengths, from zero to infinity, as there was simply no reason to have some sort of distribution. The consequence of this is that black bodies would emit an infinite amount of power, which is clearly ludicrous, and this problem was labeled the *ultra-violet catastrophe*. Planck and others were concerned about this and were trying to figure out what was going on. He came up with the idea that the atoms in the black body must be vibrating (after all, heat is just vibration of atoms), but at specific frequencies. He then saw that if the atoms were only allowed to have certain energies of vibration, that were multiples of some number (that he called a *quantum*), then he obtained excellent agreement with experiments and could predict the black-body spectrum and a whole host of other thermal phenomena that had been observed experimentally.

Planck's theory was based on the assumption that the energy due to the vibra-tion of each atom was proportional to the frequency of vibration of that atom, and the constant of proportionality is now known as *Planck's constant* (Figure 5). He published this in 1900 and obtained the Nobel Prize for it in 1918. He however only thought that this was a neat mathematical trick, and it was Einstein who real-ized that in fact this idea of things having discrete energies means that energy is *quantized* or comes in small packets called quanta. He applied this idea to light, as

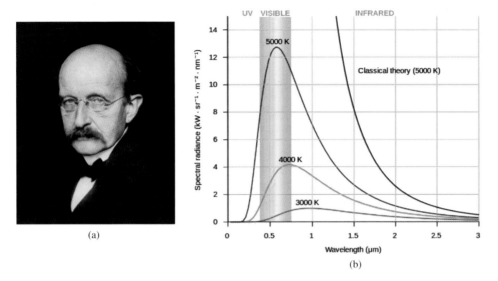

(a)

(b)

Figure 5.   (a) Max Planck in 1933; (b) Spectrum of black-body radiation, i.e. the electromagnetic radiation emitted by an ideal heated object. The red, green, and blue curves show the spectrum of objects at 3000, 4000, and 5000 K, respectively, showing that the hotter an object gets, the more it's color shifts from red to blue — by which point it will appear white. The black curve labeled "classical theory" shows the problem Planck solved — the existing theory at the time did not fit the real data at all. It deviated dramatically from observations, particularly for the visible and ultraviolet end of the spectrum, hence the name of the problem — the "ultraviolet catastrophe". Planck was able to explain the exact shape of the data and to fit it to a model that required the use of quantum mechanics.

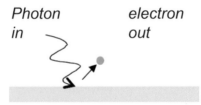

Figure 6.   The Photoelectric Effect: When light is shone on a conductor, this can cause electrons to be emitted from the surface.

at around the same time that he was working on his theory of special relativity, he was also looking at an effect in electromagnetism — the *photoelectric effect*, which makes solar cells work (Figure 6).

   This is the effect whereby when light is shone on a metal surface, under certain conditions, this leads to the emission of electrons from that surface. Using the concepts of electromagnetism at the time, it was expected that (i) increasing the intensity (brightness) of the light source would increase the energy of the electrons that are emitted — i.e. it would make them go faster and (ii) the color of the light should not matter. Instead, it was found that (i) increasing the intensity had *no effect* on the energy of the electrons, but it did lead to an increase in the *number* of electrons emitted and (ii) no electrons were emitted at all for certain colors, i.e. when the wavelength of the light was above some threshold value. When it was below this value, decreasing the wavelength (going from red to blue) led to an increase in the electron energy. He reasoned out that this meant the energy contained within the light depended inversely on the wavelength, which means it depends directly on the frequency, and that light itself comes in small parcels of energy, called photons. Increasing the brightness of a light source just means creating more photons. He found that the ratio between the energy of a photon of light and its frequency was also Planck's constant, which is a number that is characteristic of all quantum systems.

The reason for the threshold behavior is simply that electrons in any material are held there by the atoms, and in order to extract them from the material, this holding force or energy needs to be overcome. As soon as the energy of the photons in the incident light is larger than this, it has the ability to release an electron. It is also the case that one photon will lead to the release of one electron. The point then, is that even light comes in small packets of energy that we call photons, and the energy contained in a beam of light will depend on the energy of the individual photons in it, and on how many of them are present. We can therefore say that the energy associated with a beam of light can only take on certain values, as the number of photons in the beam will be an integer, and each photon will have a specific amount of energy. In physics, we call this concept *quantization*, where only discrete values (*quanta*) of energy are allowed. This extends far beyond the idea of beams of light and applies to *all* physical systems. The fact that electrons can be knocked out of metals simply by shining light on them has some interesting consequences. For instance, all spacecraft are subjected to the sun's rays, and as a direct result of the photoelectric effect, they lose electrons and end up slightly positively charged on the section that is illuminated. The imbalance of charge can lead to static electricity effects that can damage sensitive electronics. This must be taken into account when designing and building spacecraft.

Another intriguing effect as a result of the photoelectric effect is the lunar dust problem, where light from the sun that hits the moon causes some electrons to be emitted by particles on the surface, rendering them positively charged. They eventually start to repel each other and as gravity is rather weak there as compared to on Earth, they can lift off the ground *en masse*, creating a haze that can cause telescope images of the moon's surface to look fuzzy (Figure 7). The photoelectric effect is also taken advantage of for something useful in making optical sensors and night-vision equipment. As we will see later, one

Levitating Dust layer

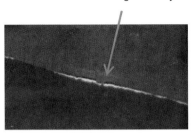

Figure 7.   Lunar dust gets charged by the photoelectric effect, as seen in this photo taken by the Surveyor 7 — a lunar lander sent to the moon in early 1968. Just above the horizon, the layer of charged dust can be seen (Images taken by Surveyor spacecraft, 1973. Published by D.R. Criswell).

could argue that the very mechanism by which we see is based on the photoelectric effect: light hits molecules in our retina, gets absorbed and via a cascade of chemical reactions, leads to the release of electrons which flow along the optic nerve to our brain.

In one deft move, we have gone from thinking about relativity and massive objects, to photons and electrons and the word "quantum". The link between the two is subtle. Electrons in atoms are often traveling very fast, in fact close to the speed of light, so relativity has a role to play in understanding how they behave. However, we do not use the theory of relativity to describe atoms, we instead use quantum mechanics, albeit with a small correction due to relativity. When Einstein came up with the idea that light comprises discrete particles called photons, he opened a can of worms that had begun back with Newton. We had better take a brief step back to the 17th century to explore that for a while.

In 1678, the wave theory of light was developed by a Dutch scientist, Christiaan Huygens. He described light as a wave motion that would behave like all other waves (for example, water or sound waves) and display the two defining characteristics of waves: interference and diffraction. Interference is when two or more waves combine to create a new pattern wherever the waves meet. One of the easiest to see examples of wave interference is a plucked string on a guitar. When the string is plucked, it sends a wave traveling along it toward both ends. At the ends, these waves get reflected and travel back, meeting each other. These waves traveling back and forth along the string then *interfere with each other*, and create the pattern we see on the guitar string, which is called a standing wave. Another example is the bands of different colors we see when we look at a CD, on the side opposite the label. This is due to light hitting the periodic structures on the CD and splitting into the different colors, via a process known as diffraction (Figure 8). Diffraction is a more complex phenomenon, and is described as the bending of light around objects — usually only noticeable very close to them.

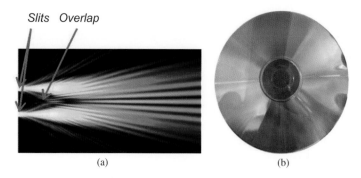

Figure 8.   (a) Interference and diffraction of light: it passes through two slits on the left and spreads out (diffracts). The light from each slit eventually overlaps at which point they interfere with each other; (b) Interference pattern obtained when white light hits a CD or DVD.

Newton came along and in 1704 published his competing, *corpuscular* theory, which proposed that light came in small packets that he called corpuscles. This was not a particularly successful theory as it was unable to explain diffraction or interference and could not account for the polarization of light. Nonetheless, Newton was held in such high regard that this theory prevailed for almost 100 years, when eventually the facts became clear — it was just wrong! The final nail in that particular coffin was hammered in by Thomas Young, who in 1803 carried out his famous *double-slit experiment*, where he passed red light through two closely-spaced slits and was able to observe an interference pattern.

Young's work, both theory and experiment, was put together with Huygens' ideas by Fresnel in 1817 to form what we now call the field of *optics*. So, the paradox arises when we realize that light is a wave and exhibits all the characteristics of waves, but that it comes in small packets, which we would otherwise call particles. Newton was *partly* right. This is the first example of the infamous *wave–particle duality*, whereby depending on the context, something can be seen to behave as either a particle *or* a wave and is the core principle of quantum mechanics (Figure 9).

This concept was proposed by Louis de Broglie in his 1924 PhD thesis. He suggested that if light, which was thought of as a wave, could behave like a particle, then perhaps particles would sometimes behave like waves. He showed that a particle would therefore have a wavelength associated with it. In fact, his idea was so revolutionary that his thesis examiners were unsure as to whether it was correct or just off the wall, so they sent it to Einstein to have a look. He approved of it, the PhD was awarded, and thus de Broglie had laid the foundations of the field known as *wave mechanics*. This later became known as quantum mechanics, combining Einstein's theory of light with Planck's ideas about quantization.

(a)   (b)

Figure 9.   (a) Louis de Broglie, who proposed the wave–particle duality; (b) A particularly potent image depicting the wave–particle duality (Illustration by Douglas Hofstadter).

These ideas were taken up very quickly by others who were researching the properties of electrons, which had only been discovered in 1897. In 1869, cathode rays were discovered by Johann Hittorf in Germany, and it was J.J. Thomson who realized through careful measurements, that these rays consisted of negatively charged particles that he called electrons. It is the flow of electrons in metal wires that we call electric current, they exert forces on each other via electric fields, they are what atoms share between each other when they form bonds, and they are what are emitted from some surfaces when light falls on them. So, electrons are pretty important. It turns out that it is actually quite easy to see their quantum nature with simple experiments. As soon as people started seeing this wave–particle duality in action, physics changed forever, and we realized that the world simply is not entirely deterministic. We still struggle with this notion and the debate continues to this day as to what exactly all of this quantum weirdness means. Many scientific lives have been spent trying to unravel this problem and reconcile what quantum mechanics tells us and how that differs from our everyday experiences. A reassuring point is that if we apply quantum mechanics to larger systems, then its predictions start to agree with those of classical mechanics, so the quantum approach always wins. We don't do this in practice as it is just too difficult, and the bottom line is that for most things, classical mechanics is *good enough.*

Now hang on a minute right there. How on earth can a solid object have a wavelength or anything else wavy about it and if this is really true, why don't we see it? The point is we do, just not directly. The electrical properties of materials, i.e. how well they conduct electricity, is a direct consequence of this waviness and is how materials conduct heat. How about some proof then? Where do we encounter waves in our everyday lives and know that they are waves? It is all to do with size, you see it really does matter — as the old saying goes. If something has dimensions from around 1mm to 1000 km, then we will notice it! The simplest example is water waves, or the waves you see in a drink in a glass or cup when it is disturbed. These waves have a typical wavelength of around 1–10 m in the sea, and around 1 cm in a cup — it depends on the depth of liquid. Everyday sound, i.e. speech and music, has a wavelength in the range 10 mm to 20 m, but as it is a disturbance of the air, we tend not to be able to actually see it. We do however make use of the wave nature of sound whenever we listen to anything in stereo, which depends on such interference (there are two speakers). The wavelength of light is also important of course, and for visible light is around 300–700 nm, with other non-visible electromagnetic radiation having wavelengths in the enormous range 10^{-12} m (gamma rays)–10^8 m (extremely low-frequency radio waves). So far, these are all things that we think of as waves anyway, so how about particles or solid objects? de Broglie showed that their wavelength was not just an intrinsic property, but depends on how fast the object is moving, which makes it even harder to understand. However, if we borrow the idea from the photoelectric effect that the energy of a photon of light depends on its wavelength, then we can say the same about a particle — it's wavelength depends on its energy. Which energy though? It turns out that the energy that matters is the kinetic energy, which is the energy contained in any moving object and depends on its speed and its mass. The relationship de Broglie deduced is that the wavelength of an object would be equal to Planck's constant divided by the momentum of the object, where momentum is the product of mass and velocity. So, to give it a value, the wavelength of an electron moving around in your mains cables is of the order 0.1 nm, which is far too small to see, hence we just don't notice it. However, as transistors and electronic components are getting ever smaller and approaching the size of only a few nm, this is starting to become noticeable. We will touch on this later also. The spacing between atoms in a typical material is around 0.1–0.2 nm, which is similar to the wavelength of electrons. Therefore, as they move around in a material (which they do, constantly), they display interference effects which manifests in the electrical properties of materials, and is ultimately why some are conductors, insulators, or semiconductors.

At that time as de Broglie was figuring this out, Davisson and Germer from the newly-formed Bell labs in New York were carrying out experiments on the interaction between beams of electrons reflected from metal surfaces. They found that when they changed the angle at which a beam of electrons hit a metal surface, the number of electrons reflected from that surface depended on the angle, in a cyclic fashion. This was a demonstration of diffraction using electrons, and to this day, electron diffraction is a commonly used tool in the characterization of materials, as it allows us to determine the arrangement of atoms. Borrowing some concepts from diffraction of light and armed with de Broglie's formula for electron wavelength, they were able to verify his wave–particle hypothesis in 1927, for which he was awarded the Nobel Prize in Physics in 1929.

Now, we are at a point where physics was developing rapidly and more accurately explaining the very big and the very small. General relativity was becoming accepted but did not find much application until the space age began where it must be taken into account when planning the trajectories of spacecraft. Quantum mechanics was being used to explain a large number of phenomena and was truly turning our understanding of how things work upside down, but there were issues that persist to today, which we will consider in a moment. The first attempt at a combination of the two theories — quantum mechanics and relativity was carried out by Paul Dirac, who devised the relativistic wave equation (known as the Dirac equation) in 1928, whereby he combined quantum mechanics with special relativity. A direct prediction of his theory was that antimatter would exist. Specifically, when looking at the quantum properties of electrons, the theory predicted that there would be particles with the same mass but the opposite charge (positive instead of negative), and that if an electron and its so-called antiparticle met, they would annihilate each other and create photons of light. A mere year later, the antiparticle of the electron, called the *positron* was discovered. The idea of antimatter as a highly dangerous substance more dangerous than any bomb has been used many times and is of course true, but thankfully it has never been isolated in anything other than single particle form which is harmless. This didn't stop Dan Brown writing Angels and Demons though — a cautionary tale about what might happen if someone actually did make a significant amount of antimatter. So, you see, the early 20th century was an exciting period indeed where we were starting to realize nature was far more complex, subtle, and rich than we had imagined. For his insight, Dirac was awarded the Nobel Prize in Physics in 1933 at the tender age of 31. An interesting character who spent much of his working life at Cambridge, he ended up retiring to Florida State University. Given the extroverted nature of quite a few of the other leading physicists of the day, Dirac appeared to be rather peculiar, although he was lucky to find love and was by all accounts very happily married for 50 years. Nonetheless, thanks to his eccentricity and that of Schrödinger, Born, Bohr, Gamow, and others at that time, there was a critical mass of theoretical and experimental work going on that accelerated the unravelling of what we now know as quantum mechanics.

So far it may sound as if all is rosy in the garden of physics and we understand everything and can model it with the appropriate mathematics. Unfortunately, this is far from the case as none of our theories, no matter how good they are, are actually complete. Undoubtedly, the biggest spanner in the works is that it became clear very early on that quantum mechanics was far from pretty. In fact, it was completely at odds with classical mechanics, at a very fundamental level. Newton's laws are what are called *deterministic*, where it is possible to predict the outcome of an experiment, or to put it another way, experiments have a definite outcome, which seems pretty obvious, and is something we don't question as it tallies with our daily experiences with the world around us. After all, if we can't repeat simple measurements then what is science all about? For example, if we hit a snooker ball at a particular speed and position, and we know about the friction between it and the cue and between it and the table, as well as the air resistance, we can predict where it will go and how fast, etc. Through experience, our brain can figure out how to hit the ball in order to achieve the desired end position. While our brain is not solving differential equations, it is using visual patterns to determine the optimal conditions and gets better with practice — this is the power of the neural network that is our brain: the ability to learn. In the quantum view of things, this is impossible, we can only predict the *probability* of the ball following a particular path. In a quantum world, the ball could follow a different path each time we hit it, and only the *average* path would be that which we expect classically. Obviously, it will be most likely to follow the classically-determined path, but there is a *finite* chance that it will not, and this becomes accentuated as things get smaller. This property of quantum systems has led to no end of discussion and angst, and even Einstein who saw very early on that this was inevitable, stated his displeasure with quantum mechanics by stating that "God does not play dice". Well, it would appear that God does indeed play dice of sorts, as the predictions of quantum mechanics have all stood up to experimental scrutiny. I have been teaching a course on applied quantum mechanics and nanotechnology for over 17 years now and have deftly managed to pass over these subtleties as they are more philosophical (and therefore impossible to answer in a satisfactory way) and do not influence how we actually apply the theory. Many books and papers have been written on this topic and how to reconcile what quantum mechanics tells us is happening and what we are used to in our everyday lives, leading to all sorts of fantastical theories involving parallel universes and the like that nobody can prove or disprove. Feynman explained it away by saying that there's *no point* in trying to understand quantum mechanics as it is totally unlike anything we know or see in our everyday lives. The important thing is to be able to apply it and learn how to interpret the results and predictions. In my own experience, this pragmatic approach works very well, and if we take a statistical view of physical phenomena, then on average (over very many experiments), quantum mechanics tends to agree with classical mechanics and our expectations. The point of course is that classical mechanics gets it wrong at quantum scales and doesn't agree with experiment at all, whereas although quantum mechanics is correct, it is far harder to use, so whenever possible, we stick with classical mechanics such as enshrined in Newton's laws. There are currently 15 different theoretical frameworks that we can use to apply quantum mechanics, but the most commonly used and accepted is that which was developed by Niels Bohr and Werner Heisenberg in the mid-1920s, while working in Copenhagen. This is now known as the Copenhagen interpretation of quantum mechanics and is the approach taken in this book to explain phenomena. This is the only branch of physics where a theory is widely used, is mathematically rigorous, but is also full of paradoxes and frankly just doesn't seem right. It does however agree with experiment which after all is what it's all about!

The devil is in the detail here, as individual experiments at quantum scales can vary wildly. This should not be taken as a way of saying quantum mechanics is random, as it certainly is not. It is more that there are simply deeper-level patterns that we do not fully understand yet. Feynman himself managed to combine quantum mechanics with electromagnetism, coming up with a theory that describes how light interacts with matter, called *quantum electrodynamics*. To date, this is the most accurate theory in physics, with agreement between theory and experiment having been verified to an enormous degree of precision — around 10 parts per billion. It does not explain why particles have mass however, and this is something that was only answered with the proposal and subsequent discovery of the Higgs boson. In the meantime, quantum mechanics has been combined with special relativity in more detail, resulting in quantum field theory, but to date, there has been no success in combining quantum mechanics with general relativity apart from some very special cases involving black holes — Stephen Hawking's contribution. This was Einstein's last quest and has been attempted by many a frustrated physicist. It may be that such a grand unified theory as it is known, simply does not exist. However, it does make sense to at least believe that there is a theory of everything, to which quantum mechanics, general relativity, special relativity, Newton's laws, and electromagnetism are mere approximations. For now, though, we have to settle with each of these separate theories and hope that we continue to use them appropriately, whilst being aware that they are not complete. A good physicist knows not only about each theory, but when they can and cannot be applied.

Nanotechnology is based on the fact that the properties of very small (nanometer scale) things are different from bigger things, and in ways that are useful to us. This is largely due to quantum mechanics. For this reason, we should consider a number of key messages of quantum mechanics, so that when we are looking at nanotechnology we will have a better understanding of why things are the way they are, and we will also have an idea of the limits of our understanding.

To get a true appreciation of just how quirky quantum mechanics is, we need look no further than Young's double-slit experiment for light, but now

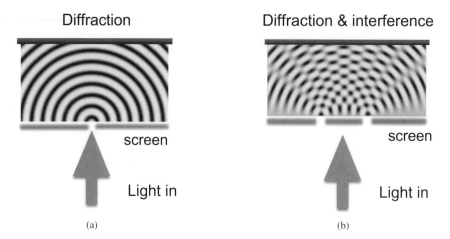

Figure 10.   (a) Diffraction pattern obtained whenever any wave (water, sound, light, electron) is passed through a thin slit whose size is comparable to the wavelength; (b) Interference pattern obtained whenever any wave (water, sound, light, electron) is passed through two slits whose size and separation is comparable to the wavelength.

repeat it with electrons. First, though, if we shine light (of one color, so with one distinct wavelength) onto a screen with a tiny hole in it, it spreads out as shown in Figure 10(a).

This spreading out pattern, known as diffraction, is a characteristic feature of all waves, another example is shown in Figure 10(b), in this case a real wave produced when something is dropped into water. If we instead take a screen with *two* holes that are close together, we end up with an *interference* pattern as shown on the right. This is the sort of pattern you get when any two waves interact and is seen with sound waves and water waves as well. If we were to place a camera at the top, it will see a series of bright and dark patches called *fringes*. The key points are that there are some bright areas and some dark areas that are arranged in a regular pattern, and it is straightforward to describe this using simple maths. The bright and dark areas correspond to regions where there is *constructive* and *destructive* interference, where the two waves add together or cancel each other out, respectively.

Now, let us repeat the experiment with spray paint instead of light. We will use a can of spray paint, a metal sheet with a hole in it and a nice clean wall. In that case then, we will have a situation like that shown in Figure 11. If we were to look at the wall, the paint pattern would look as shown in the figure.

Now, if we add an extra hole to the metal plate, we will end up with two paint spots. This is what we would expect and is what we see if we do such an experiment. Things become far more complicated in the quantum world. Imagine we

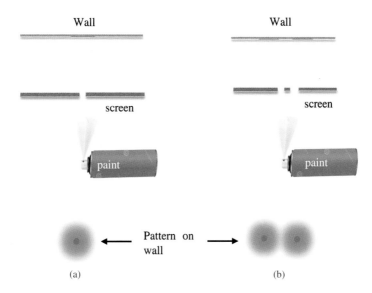

Figure 11.   When particles pass through small holes — the resulting patterns when paint is sprayed through a hole in a screen. (a) Pattern on wall for one hole; (b) Pattern on wall for two holes.

could make the paint droplets smaller and smaller, approaching the size of atoms or even smaller: a fraction of a nanometer in diameter, and we could make the holes smaller and closer together, then the pattern that we would see on the back wall will evolve from the well-defined paint splatter to the wave-type pattern we saw on the last page. If we were to measure the spacing between the "peaks" of the pattern, we would find that they show the paint particles have a *wavelength* as predicted by de Broglie. This *wave nature* of particles, particularly noticeable for electrons, is taken advantage of in the electron microscope (which was invented in the late 1920s), which focuses a beam of electrons down onto a sample in much the same way as an optical microscope focuses light. The limit to resolution (i.e. the smallest spacing between two objects that can be clearly seen) in an optical microscope depends on the wavelength of light we are using, and for normal white light, where the wavelengths are 300–700 nm, is around 300 nm. In an electron microscope, the wavelength of the electrons is far smaller, at around 0.01 nm, leading to a resolution of the order 0.1 nm when everything is optimum.

When dealing with waves, the arrangement of a diffraction pattern depends on both the wavelength of the waves and the spacing of the holes in the screen, and a simple formula relates the two. This is used in the technique known as electron diffraction, which has been used for over 80 years to measure the spacing between atoms in materials by shining a beam of electrons through a thin film of the material of interest and looking at the pattern the beam diffracts into. A detailed knowledge of the arrangement of atoms in materials is important, as it allows us to understand why materials have the properties they do, and hence to design new ones for specific applications. Unfortunately, quantum mechanics gets far more bizarre and beggars belief sometimes. Let us consider carrying out the diffraction experiment again, with either light or electrons, but when the brightness of the light source or the electric current of the electron beam is turned down to be low enough that only one photon or electron is traveling through the system at a time. We can then consider that photon or electron to be a particle rather than a wave. We would expect the pattern of light or electrons hitting the back wall to look like the paint splatter, because after all, how can we expect to see a diffraction pattern if there is only one "particle", and therefore nothing for it to interfere with?

If only it were so — in reality, we find that if we wait long enough to be able to see the pattern the photons or electrons made when they arrived, they follow the diffraction pattern we obtain when we have waves. This means that the "particle" was really a wave and *it interfered with itself* (Figure 12). How? This is where physicists get sheepish and start to look shifty and admit that we have absolutely no idea. We can describe it very simply using mathematics and

Double-slit experiment with electrons

(a)   (b)

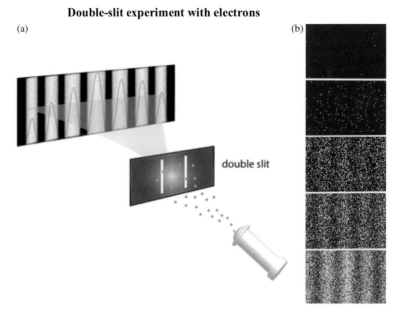

double slit

Figure 12.   (a) Schematic of the double-slit experiment with electrons. A beam of electrons is passed through two small slits and form an interference pattern — even when only one electron passes through at a time; (b) Experimental observation of this interference pattern. The sequence of images were taken at different times. Note that we can see bright spots which indicate where individual electrons have arrived (particles) at the same time as the bands which indicate interference (waves) (*Image courtesy*: Dr Tonomura).

perform all sorts of calculations and predictions that turn out to be true, but we stand no chance of really understanding it as *it makes no sense*.

It gets even more strange if we try to measure *which* hole the electron or photon went through. If we do so, then we get the paint splatter pattern again. The reason for this is that in the process of measuring where the electron or photon went, we have to disturb it somehow, either by trying to shine a light off it, or by measuring its electric or magnetic field. This has the effect of disturbing its quantum wave and destroys its ability to interfere with itself. Nobody has ever been able to devise any measurement that is gentle enough not to cause this, and there are good reasons to think that it is impossible to do so.

So, we get it — nature is just weird, and our understanding of the universe is limited to the sort of things we see in our everyday lives. Thankfully, this is good enough for most things, but due to our inquisitive nature, we will always want to know more.

There are a number of examples of everyday things that quantum mechanics can explain, and we should have a look at them, as they will pave the way toward appreciating all things nano. The first example we will briefly consider is radioactive decay. There are three mechanisms by which atoms can decay radioactively: alpha, beta, and gamma decay.

Alpha decay, as discovered by Rutherford in 1899, is when an unstable nucleus emits a so-called alpha particle, which is the same as the Helium nucleus, i.e. it consists of two protons and two neutrons. An example is the decay of Uranium-238 to Thorium-234. It was known that whenever an atom underwent alpha decay there was a relationship between the energy of the emitted alpha particle and the half-life (i.e. the time taken for half the atoms in a piece of radioactive material to decay) of the material, but it was not known *why* (Figure 13). What *was* known though was that there was a force keeping the alpha particle within the nucleus, and that

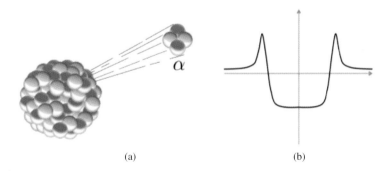

(a)   (b)

Figure 13.   Alpha (α) decay was one of the first processes to be explained by quantum mechanics. (a) A nucleus decaying by emitting an alpha particle; (b) The binding energy of an alpha particle within a nucleus, as a function of distance from the center.

sporadically one would pop out. Through detailed measurements, Rutherford had a good idea of the nature of that force, which we call the *strong nuclear force*, and how it varied as a function of distance away from the nucleus. It wasn't clear how an alpha particle could possibly overcome this force until Hund started looking at the idea of quantum tunneling, which Gamow then applied to solve the relationship between half-life and alpha particle energy, wherein the problem was solved.

Quantum tunneling is best described by analogy. We all know that if we take a ball and place it on the side of a hill, it will roll down. We also know that it will not spontaneously roll uphill. Why is this? Well, it rolls downhill to reduce its potential energy, which is a general law of nature for some reason that we don't understand either. To roll uphill would require energy to be added to it, and it seems that energy is conserved, so this doesn't happen. In the quantum universe, there is a chance that it might suddenly appear on the other side of the hill, as if it has popped along a tunnel through the hill (hence the name *tunneling*) (Figure 14). We can argue that there is a chance it might gain enough energy for a short enough time to roll over the top and appear on the other side, which is a way of saying that energy is conserved *on average*, but it can fluctuate wildly for very short periods of time. Both explanations for tunneling don't actually make any sense, but the fact remains that quantum tunneling does happen, and there are quite a few examples of it. It is particularly relevant to nanotechnology, as the distance over which tunneling occurs is typically of the order 1 nm, so we really notice it in nanosized things.

These fluctuations in energy are described by the *Heisenberg Uncertainty Principle*, which is a direct consequence of the wave nature of things. In its more commonly known form, the uncertainty principle tells us that there is a limit to the precision with which we can know both the position and momentum of any object

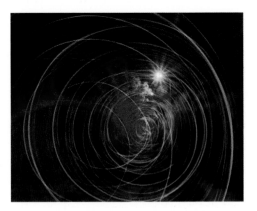

Figure 14. According to quantum mechanics, particles can overcome potential barriers by tunneling through them.

at the same time. This might sound like an innocuous statement, but it has far-reaching consequences. For example, you would imagine that if something is stationary, then you could measure its position perfectly, and then be able to say that its momentum is zero. The uncertainty principle tells us that this *cannot* be the case, that nothing can be completely at rest, it must be moving a little bit, so that it always has some momentum and there's always some uncertainty in its position as it jiggles about randomly. We do not see this in our everyday lives, as the fluctuations in position and momentum are so small, but they are very noticeable down at the size scale of atoms. The corresponding spontaneous fluctuations in energy that constantly exist at this scale may even have been responsible for kick-starting the big bang. This also means that even at the lowest possible theoretical temperature, 0 Kelvin, atoms in materials are still moving about a bit — completely randomly. This is due to something called *zero-point energy* and is something that many people have tried to tap into, ignoring the fact that thermodynamics tell us that to do so is unfortunately not possible.

There are many more examples of quantum tunneling, including a variety of electronic components that use it to their advantage, and of course the scanning tunneling microscope, the invention of which many would agree started the whole field of nanotechnology. Tunneling is also problematic in some situations, for example, the transistors in microprocessors are now getting so small that quantum tunneling is starting to become a nuisance, as some electric current is now going places we don't want it to. On a completely unrelated note, there is some evidence that quantum tunneling of protons leads to distortion of the bonds that hold atoms together in DNA, which in turn, leads to a change in its structure, which is otherwise known as *mutation*. There is a possibility that this is a contributing factor to cancer, although the evidence is by no means clear. Nonetheless, this is something that warrants further research as it suggests that where cancer is concerned, it may occasionally occur spontaneously and does not necessarily always have to be triggered by something external.

The uncertainty principle is often explained with reference to the thought-experiment dreamt up by Schrödinger, one of the founders of the theory of quantum mechanics as we use it today; that we know as *Schrödinger's cat* (Figure 15(b)). This is a very subtle idea encompassing uncertainty, superposition, and entanglement, all concepts that we will not delve any deeper into but are central to the operation of quantum computers.

This experiment is essentially the same as our electron or photon passing through a hole in a screen — the pattern it forms on the wall makes it impossible to determine which hole it passed through, and it behaves as if it went through both (hence the interference), unless we try to measure which hole it has actually gone through in which case the interference disappears. Schrödinger's analogy

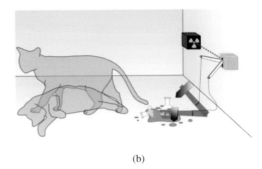

(a)   (b)

Figure 15.   Werner Heisenberg was one of the early physicists who devised quantum mechanics. (a) Heisenberg in 1933; (b) Depiction of Schrödinger's cat — the thought experiment he devised to highlight the principle of superposition in quantum mechanics. A hypothetical cat is placed in a sealed box containing a vial of a radioactive substance and a hammer. At a randomly chosen time, this hammer smashes open the vial, and the cat is exposed to deadly radiation. Exactly when this happens is unknown. In the framework of quantum mechanics, we must describe the cat as having a chance of being alive and a chance of being dead, until we open the box and have a look. This state of possibly being alive and dead at the same time until we look is called superposition. From a practical point of view, this argument makes no sense as the cat is either alive **or** dead — it cannot be both at once. However, we need to assume it is in this superposition to agree with the findings of quantum mechanics.

was that there was a cat placed in an opaque box with a radioactive source. This source could decay at any moment (randomly), killing the cat by radiation poisoning, but we have no way of knowing when this might happen, or indeed *if* it has happened until we open the box to take a look. From a common-sense perspective, we would say the cat is either alive or dead. Using the ideas of quantum mechanics, until we open that box we don't know which it is, so we would mathematically describe the cat as being both alive and dead, as there is a 50:50 probability of both being true. We would write a formula that basically says probability of cat being alive is 50%, and probability of it being dead is 50%. Therefore, the total probability of the cat being either alive or dead is 50% + 50% = 100%. Following our observation, we could then say either the certainty of the cat being alive is 0% or 100%, so we could argue that our measurement has changed the probabilities, where in fact all it has done is convert probabilities into certainties. The same happens in quantum systems but is more than just mathematical gymnastics. What we have come to learn over the last 100 years is that simply by doing a measurement on a quantum system, we actually change the system's behavior. This sort of subtle and somewhat confusing reasoning is part and parcel of quantum mechanics, and the dramatic effect that simple observations have is one of the reasons why it makes our head hurt!

Quantum Biology

Another example of quantum mechanics at play is in photosynthesis: the process whereby plants use chlorophyll (the substance that makes plants look green) and sunlight to turn carbon dioxide and water into glucose and oxygen. When chlorophyll absorbs sunlight, it uses that energy to release an electron, which travels, via a number of molecules, to combine with the complex molecule NADP, grabbing a hydrogen atom from a nearby water molecule, releasing the oxygen left over. This then leads to a molecule, NADPH, which reacts with carbon dioxide and water to produce carbohydrates, culminating in glucose.

When the photon and electron that are involved in photosynthesis are traveling along their pathways to the chlorophyll molecule and the NADP, respectively, they see a complicated energy landscape involving a series of energy barriers that need to be overcome. This can happen via quantum tunneling, significantly increasing the efficiency of the overall process. As the electrons flow along these molecular pathways, they can be thought of as waves, sampling all the pathways in their vicinity, probing to see which is the most efficient, and then going that way. This is not a conscious effort, it is simply that the electrons flow, with wave-like characteristics, they spread out in accordance with the uncertainty principle (uncertainty increases with time) and are most likely to follow the path of least resistance, which just *looks* as though they are deciding which way to go. How do we know this? Well, when we measure how quickly the electrons travel through this system, we can only explain that using the above concept. If they simply flowed as electrons without having this quantum sense, the whole process would be orders of magnitude slower.

Recent experiments have tried to explain our sense of smell as being quantum in origin. For many years, it was believed that our sense of smell was due to a lock and key mechanism at play in the olfactory sensors, i.e. that a particular sensor was triggered (i.e. creates an electrical impulse that our brain interprets as a smell) if a molecule of the right shape landed on it. Recent evidence has pointed to the fact that it is actually the *vibrations* of molecules that we are sensitive to rather than their specific shape. Molecules of exactly the same shape but made using different atoms, and therefore having different strength bonds which vibrate at different frequencies are known to smell differently. The nose appears to use

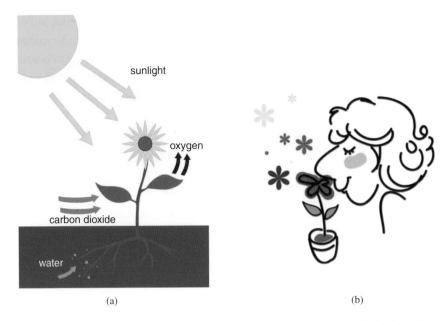

Figure 16.   (a) Photosynthesis is an example of quantum mechanics at work; (b) Evidence is mounting that the sense of smell is due to quantum tunneling.

quantum tunneling to measure these vibrations — so our sense of smell uses the transfer of electrons from the molecule of interest to the molecular sensor in the nose, and the fact that if a molecule is vibrating, this changes the energy of the tunneling electron a little bit (Figure 16). The jury is still out on this one as there is not quite enough experimental evidence for this hypothesis to make it a done deal just yet. This and photosynthesis are only two of many examples of quantum mechanics having a role in biology, and the field of quantum biology, as first proposed by Schrödinger in 1944 is striving to find others.

The point therefore, is that modern physics is deeply subtle and complex, it deals with phenomena that we do not normally see in our everyday lives, but as soon as you scratch beneath the surface, you find that actually, it's everywhere. We are in that middle ground where our experience usually deals with things too small for general relativity to be relevant and too large for quantum mechanics to explain, but every now and again, we see both at play. Although quantum effects tend to occur over very short distances or times, the consequences can be felt over much larger distances, large enough for us to notice.

Seeing the Light —Size Matters Yet Again!

Even the process of how we see — how our eyes convert light into an electrical signal relies on quantum mechanics, and is a bit like the photoelectric effect,

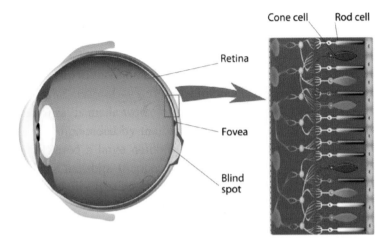

Figure 17.   Structure of the mammalian eye showing the "pixels", or rods and cones — molecules that can turn light into electrical signals.

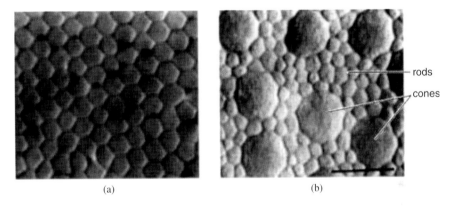

Figure 18.   The retina. (a) An image taken near the fovea, where the cones are at their highest density; (b) Away from the fovea, where there are cones and rods. The scale bar on the right is 10 microns. If rods were considered to be pixels, there are approx. 12,700 pixels per inch.

except at the molecular level (Figure 17). Within the rods and cones — the light detectors that make up the retina, there are molecules that absorb light (there are different cones for red, green, and blue, and the rods absorb over a wide range of colors apart from red), and when they do, the energy in the light leads to emission of an electron from those molecules, which is then carried away to the brain as an electrical signal by the optic nerve (Figure 18). There are around 5 million cones centered in an area approximately 300 microns (0.3 millimeters) across at the macula, corresponding to a linear density of around 12,700 cones per inch. The cones at their end are around 130 nm across, although they can be several tens of microns long — they are packed a bit like spaghetti in a jar. The rods are even narrower at less than 100 nm across.

By comparison, the highest resolution display on a smartphone has around 315 pixels per cm or 801 pixels per inch, translating to a pixel size of around 30 microns, so 300 times bigger than in the eye. A display has been demonstrated using OLEDs (organic light emitting diodes) with a pixel size of the order 5 microns, although this is not yet in production, and to be honest (i) it's not clear that the eye would be able to see the difference between pixel sizes of 50 microns and 5 microns and (ii) the graphics processor would have to be insanely powerful. The latest iPhone (yes, I'm a bit of an Apple fanatic) has approximately 2 million pixels at 401 pixels per inch. A similar display size with 5-micron sized pixels would have in excess of 200 million pixels. Although these are all far larger than the size of our rods or cones, they are small enough that we can't detect that they are discrete (Figure 19). Still, calling them a "retina" display is a bit cheeky!

On the subject of light, one of the triumphs of quantum mechanics in the applied sense was the invention of the laser in the late 1950s by Townes, Schawlow, and others. The basic principles behind the operation of the laser had been laid down by none other than Einstein himself back in 1917. The history of the invention of the laser is akin to that of the discovery of the structure of DNA — several independent groups developed the idea at around the same time, and fought about who was really first for decades afterwards — the controversy continues. I mention the two names above as they were awarded the Nobel Prize for their efforts. I was fortunate enough to attend a public symposium given by Schawlow and four other Nobel laureates in science and medicine when I was still an undergraduate at Trinity College Dublin, and I remember he was given a

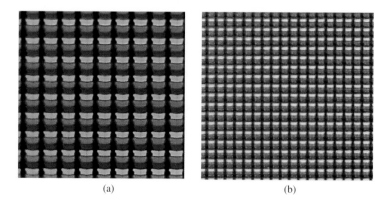

(a)   (b)

Figure 19.   Pixels. (a) The pixels in an iMac display, with 100 pixels per inch resolution; (b) The pixels in an iPhone 5S, with 326 pixels per inch — impressive, but still 40 times larger than the detectors in the mammalian eye.

standing ovation when asked what he did with his share of the prize money. The others had mostly used it to pay off their mortgages, whereas Schawlow used his share to help set up a home for autistic children in California (one of his three children was autistic), a choice which has a deep personal resonance for me having seen first-hand what a devastating impact this condition has on children and families.

The laser has become ubiquitous since the 1970s when diode lasers became cheap enough to be rolled out for a range of applications. We now use them in barcode readers in retail outlets, for aircraft navigation, for range finding, measuring room sizes, remote sensing, pointers for giving presentations; and their relatives — light emitting diodes are used for all sorts of applications, including remote control handsets, displays, including phone screens, TVs, and computer monitors (Figure 20). The principle of operation of laser diodes relies entirely on quantum mechanics, and the active "lasing" region where the light is created is often only a few hundred nanometers across.

The color and appearance of things depend very strongly on both their constituent materials and their size. We all know that clouds are white, even though they are made of water which we know to be transparent. The color is due to the fact that the water is in droplet form, and there are so many droplets (their typical diameter is around 10–20 microns) that there are lots of reflections of light from the droplet surfaces. As a result, light that enters a cloud gets scattered all over the place, which makes it appear white. A similar effect happens in milk — it comprises an "emulsion" of particles surrounded by water (Figure 21). These particles largely come in two forms: fat globules that can vary in size from around 1–20 microns, and casein micelles that are typically from 1 micron down to around 100 nm across. The fat globules do not mix with water for two

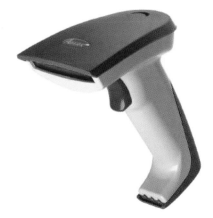

Figure 20. **A Barcode Reader** — One of the everyday applications of lasers.

Micelle

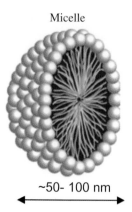

~50- 100 nm

Figure 21.   An emulsion, e.g. milk, contains micelles that have the water-hating (hydrophobic) ends on the outside. This is an illustration of a micelle cut in half.

reasons — first, they are less dense than water so will tend to float and second, they are *hydrophobic*, a term which literally translated means "fear of water".

Water is *polar*, which means that water molecules have a slight difference in electrical charge between their ends, with the result that water mixes very well with other polar materials. Fat on the other hand is non-polar, so there is no attractive force between it and water, so they effectively repel each other. If fat is added to water and given a good shake however, it breaks up into tiny particles which remain suspended, which is what we call an *emulsion*. These particles are of a similar size to the water droplets in clouds, and therefore, for the same reason as above, milk appears white. However, if we compare milk with different levels of fat content, so look at full-fat milk (3.5% fat by weight), semi-skimmed (1.5–1.8% fat) and then skimmed milk (less than 0.3%), there is a difference in color — skimmed milk has a slightly bluish tint relative to the others. This is due to the fact that there are very few fat globules, so the light is predominantly scattered by the much smaller casein micelles which scatter blue light very effectively.

This issue of size making a difference to the appearance of something is important enough that we should consider its origin. When we look at an object, what our eyes are doing is collecting light reflected from it and focusing it on our retina, which converts that into a series of electrical signals (one from each detector, i.e. rods and cones), in much the same way as a CCD or digital camera works with pixels taking the place of the rods and cones. Our brain then decodes those electrical signals and uses them to reconstruct the object, containing information about position, size, shape, color, and finish (i.e. whether the surfaces are matt or shiny). This is based on the fact that light bounces off things in a well-defined way. However, when objects become small enough, and particularly when their size is comparable to the wavelength of light (yes, in the nanometer range), this process becomes disrupted and other effects such as diffraction start to come into play and be very noticeable.

The scattering of light off raindrops and fat globules in milk (and in fact, anything spherical) was described by Gustav Mie in 1908. Mie used Maxwell's equations (the four equations that describe everything to do with electric and magnetic fields and electromagnetic waves, excluding any quantum effects) to calculate what happens to light that is incident on a sphere. He was able to show that scattering was strongly dependent on particle size: certain sizes scatter much more strongly than others, depending on the wavelength of light used. This was used successfully to explain the color of clouds and milk as well as many other related phenomena, particularly where the particle size is much larger than the wavelength of light. A special case where the particle size is smaller than the wavelength of light (a regime known as "Rayleigh scattering") is particularly interesting, and accounts for the bluish tint of skimmed milk, as well as the fact that the sky appears blue (Figure 22). To see how this works, consider a small sphere, whose size is comparable to the wavelength of visible light: when we shine light on it, it scatters that light off in all directions, but with a well-defined distribution. The angle by which light scatters depends very strongly (to the fourth power) on the wavelength, the most scattering occurring for short wavelengths — the blue end of the light spectrum (Figure 23). In other words, when you shine light on small particles, the longer wavelengths (red) carry on in the same direction they were going, whereas blue light goes off at a right-angle. This happens in the atmosphere, when light scatters off molecules, dust particles, and smoke, making the color of sunlight different depending on which direction you look. This can be seen very nicely just before sunset and

(a)   (b)

Figure 22.   Objects such as clouds and milk appear white due to the presence of particles in them. (a) Clouds appear white as the water droplets in them are between 1–100 microns in diameter, and scatter light uniformly in all directions; (b) Milk appears white as the fat globules droplets in it are between 1—100 microns in diameter, and scatter light uniformly in all directions.

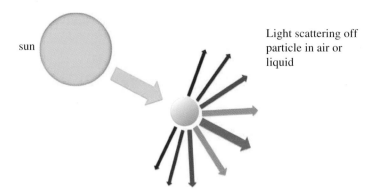

sun

Light scattering off
particle in air or
liquid

Figure 23.   Scattering of light from small particles. The shorter the wavelength, the more light gets scattered. The figure shows sunlight striking a particle or molecule in the atmosphere. The longer wavelengths toward the red end of the spectrum are barely scattered, so continue mostly unaffected, whereas the shorter wavelengths, toward the blue & violet end are strongly scattered. A person standing under the particles would see it as blue, whereas a person standing in line with it and the sun would see it as red.

after sunrise — if you look toward the sun, it appears orange/red, and as you turn your gaze upward, the sky appears more and more blue. The reason why it is particularly pronounced at these times is because the sun's light is glancing along parallel to the earth's surface coming from near the horizon and has to pass through more of the atmosphere than earlier in the day when it is just passing straight down through the atmosphere. This longer path leads to more scattering and therefore a more noticeable effect (Figure 24).

A similar effect is noticed with shaving gel — it starts off blue/green, and then turns white. This happens as it contains a foaming agent which becomes active upon contact with air — so it forms lots of tiny (around 10 microns in diameter) bubbles, which scatter light in all directions, hence the white appearance.

This variation in scattering versus size for water and fat droplets is partly due to the wave-nature of light — if the length of the path the light takes through the particle coincides with an integer number of half-wavelengths, then a standing wave is set up inside the particle, and there is very little scattering, and most of the light just passes straight through. However, when the particle is made from metal, when the light hits it, it makes the electrons in the particle shake back and forth. If the electrons do this easily, and the particle is small compared to the wavelength, then at any one instant, they all "see" the same light fields, and they all respond in the same way and move together. This combined motion of electrons is called a *plasmon*, and leads to greatly enhanced absorption, which in turn, depends very-strongly on the color of the light. This has the result that if we shine white light on a small metal particle, because the different wavelengths of light will interact

Figure 24.   An example of scattering by particles in the atmosphere (molecules, dust & water droplets) at sunset creating a red–orange glow. The sky in the foreground is dark blue.

differently with that particle, the light that we see will no longer be white, but will
be colored. We can tailor that color by changing the size of the particle, or by
changing the metal that it is made out of, as different metals will have a different
response to light, i.e. different wavelengths at which plasmons are created.

This effect is exactly what we referred to in the introduction when we talked
about stained-glass windows and the Lycurgus cup. Both used gold and/or silver
nanoparticles known as *colloids*. The first person to realize that the color of nanoparticles depends on size was Michael Faraday, who performed experiments on
gold nanoparticles suspended in solution. In Figure 25, I have shown the color that
we see for suspensions of gold (Au) and silver (Ag) nanoparticles as a function of
their size from 300 nm down to 50 nm.

One of the solutions Faraday prepared is currently on display at the Royal
Institution, where he had his lab and carried out much of his groundbreaking work
on electromagnetism. Faraday is widely considered as the first researcher in nanoscience and nanotechnology. He noticed that suspensions of gold nanoparticles
had a different color than bulk gold, and that when light was shone through such
suspensions, the beam of light was visible while in the liquid. This phenomenon
is known as the *Faraday–Tyndall effect*. When artists were making gold-based
colors, they simply had to change the way in which they stirred the mixture, and

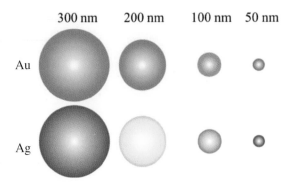

Figure 25.   How the color of Gold (Au) and Silver (Ag) nanoparticles varies with size for diameters in the range 300–50 nm. This effect has been used in art for millennia.

Figure 26.   (a) The colors in a peacock's wings are created by naturally-occurring nanostructures; (b) A zoom-in on a peacock feather showing the periodic structures that give rise to the blue color.

in this way, they could change the color. The paints used nanoparticles that were below 100 nm across, and the stirring process caused them to stick together. The longer one stirs for, the larger the particles get, evolving from blue to red in color as they get bigger. Of course, the reason for this was totally unknown at the time and has only been elucidated in recent years with our increased understanding of plasmons and advanced imaging capability. The field of science that deals with the interaction of light with charges in materials is known as *plasmonics* and has led to a large number of technological breakthroughs, including more efficient laser diodes such as those used in CD and DVD players and a variety of very sensitive and precise materials analysis tools. There are even examples in nature of animals using some of the ways that light interacts with nanostructures to enhance their color or brightness — the wings of butterflies and peacocks are examples of this (Figure 26).

Figure 27 shows a *Morpho didius* butterfly. This is a perfect example of struc-tural iridescence — the blue color is not as a result of any pigmentation, it is

(a)

(b)

Figure 27. (a) A Morpho didius butterfly. The blue iridescence is due to the presence of periodic structures on the wings; (b) An Electron microscope image that shows these structures on the butterfly wings — they have the same spacing as the wavelength of blue light, so they scatter it most effectively and therefore appear blue.

instead due to periodic structures on scales all over the surface of the wings, as shown in the electron micrograph. The spacing of these ridges is the same as the wavelength of blue light, so the wings appear blue. For comparison, the scales that cover a moth's wings are arranged in a random fashion, which leads to moths having dull wings, without any bright colors.

Structural iridescence has been created artificially, and there is now a range of paints that contain a variety of nanoparticles, mostly titanium dioxide, silicon dioxide or alumina or a combination of these, and as their size is comparable to the wavelength of visible light, their apparent color depends on the angle at which they are observed. This is sometimes called pearlescence or goniochroism. An example is shown in Figure 28 — a wall with a pearlescent paint coating that looks red when viewed straight on, and green when viewed at an angle.

Figure 28.   A wall with a pearlescent coating that appears to have a different color depending on the angle it is viewed at.

Now that we have considered the basic principles behind the behavior of nanosystems, we should look at specific examples of materials that make use of the novel properties of nanostructures — i.e. nanomaterials. We will see that there is an enormous number of applications of these and I hope you recognize a few everyday ones as you go along.

Chapter 4

Nanomaterials

Discovering the secrets of the Universe is one thing; ensuring that those secrets are used wisely and appropriately is quite another — Attributed to Prince Charles in the *Independent on Sunday*, which led to several UK newspapers scaremongering with headlines referring to his fear about "gray goo" — a claim that Charles later denied.

    I wish I had never used the term "gray goo" — K. Eric Drexler, author of *Engines of Creation* and the first person to predict self-replicating nanorobots — which he called gray goo (Figure 1).

 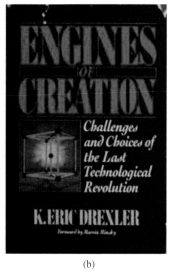

(a)   (b)

Figure 1. (a) Prince Charles who commissioned the 2003 Royal Society report into Nanotechnology; (b) The book that inadvertently started the gray goo story.

Nanomaterials — What They Are & Why We Use Them

Materials are the matter or substance(s) from which things are made. The raison d'être of nanotechnology is the fact that many materials, when small enough, possess different and advantageous properties than larger pieces of the same materials, and by small enough I mean nanometer-sized. The properties of materials such as strength, toughness, electrical conductivity, magnetic behavior, color etc., all have characteristic *lengthscales* associated with them and so their size is all-important. These lengthscales are typically in the range 1–100 nm.

For instance, the electrical conductivity of a metal is associated with a quantity called the mean-free path. This is how far an electron can travel on average through a material before it hits something. Electrons are surprisingly clumsy when they flow through materials (this flow is what we call an electric current). They hit other electrons, atoms, surfaces, and any defects that are present — all due to the wave nature of electrons. The typical value of mean free path in a metal is of the order 40–50 nm, and as soon as we make a conductor smaller than this, its electrical conductivity starts to change and dramatically depend on size. With mechanical properties, the strength of materials depends very strongly on their microstructure, i.e. how grainy they are, and on the average distance between defects — another important lengthscale. Another lengthscale that matters is the average size of the grains in a metal. This varies widely but is typically in the range 10 nm–10 microns. It has been shown that when making something out of metal, its strength increases as soon as its size drops below that average grain size. Just by changing the number and types of grains (this is collectively called the "microstructure"), we can influence the mechanical properties of a metallic material significantly. To understand why, consider what happens when we hit anything. Let's take the specific example of a hammer, as shown in Figure 2. When you hammer a nail into a wall, you want the momentum within the hammer to be transmitted to the nail. What we do not want is the energy from the impact to be simply absorbed by the hammer.

The same is true of golf clubs — we want the energy from the swing to be transferred into the golf ball rather than be absorbed within the club (or indeed the ball). The underlying principle is that when we hit a material with grains, energy from the impact is used up in making the grains slide against each other. This energy then gets dissipated as heat and is wasted. How about a material without any grains? There are two ways to achieve this — either a single crystal or an amorphous (disordered) or glassy atomic arrangement. The first option is not ideal as crystals tend to be rather brittle and the manufacturing effort, complexity and therefore cost required to create large-enough crystals is cost-prohibitive. The second

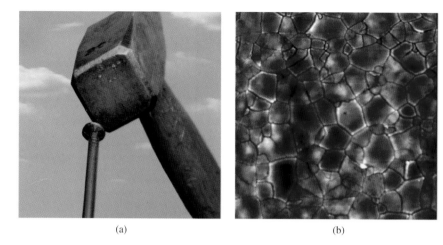

(a)   (b)

Figure 2. (a) A hammer striking a nail; (b) Grains in a metal. This is an image of a region 10 microns across showing that grains in this material are around 1 micron in width. When struck, the grains move against each other and generate heat, so the impact gets dissipated.

option is what we have in practice. The microstructure or lack thereof is designed with the application in mind, and the specific materials are chosen accordingly.

These are just a few examples, and as we have seen, the color of materials can change when they are in the nm size range due to quantum effects. The relevant lengthscales of materials within which we start to see size-dependent properties are shown in Figure 3 from which we can see why the whole field of nanotechnology even exists. The influence of nanotechnology extends into quantum and classical mechanics, but also touches on large-scale and ultra-small-scale systems related to relativity and quantum field theory. The ultimate limit to size is at the so-called *Planck Lengthscale*, which is around 10^{-35} m. This is considered to be the size of superstrings or superloops, or whatever it is exactly that makes up spacetime. We do know that spacetime is not continuous and smooth at these lengthscales and has some graininess.

The boundaries between the different regimes such as quantum and classical mechanics are not clearly defined, so which framework we use depends very much on the particular problem we are looking at. The most important rule of thumb is that if the size of an object is comparable to its quantum wavelength, then quantum mechanics makes its presence felt. For example, the de Broglie wavelength of a typical electron in a metal is of order 0.5 nm. On its own, this is unremarkable. However, it is comparable to the spacing between atoms, which in most materials is in the range 0.1–0.2 nm, so the wave-nature of the electrons is noticeable. It is responsible for the band structure of materials, which determines whether they will be a conductor, semiconductor, or insulator.

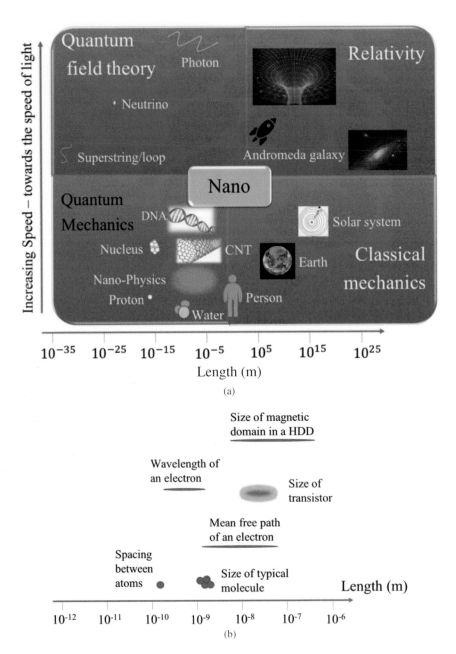

Figure 3.   (a) Lengthscales in physics. For very large or quickly-moving (close to the speed of light) systems, relativity is the appropriate framework to use; for very small and quickly-moving, we have quantum field theory; for everyday sizes up to around 10^{15} m and slow speeds, we have classical mechanics and for small, slowly-moving systems we have quantum mechanics. Nanotechnology bridges all of these areas, but mostly the regime between quantum and classical; (b) Characteristic lengths associated with materials.

The ability to control material properties at ever-smaller scales has led to developments starting with improved steels toward the end of the 19th century, and many of the alloys, composites, and advanced materials in common use today. Many of these incorporate nanomaterials in some form or another.

The definition of a nanomaterial is that it is any material that has at least one of its dimensions below 100 nm, although in some cases, that is extended to 1000 nm (1 micron). The definition of a nanomaterial that the European Commission adopted in 2011 reads as follows, in typically baffling EC lingo:

> *A Nanomaterial is a natural, incidental or manufactured material containing particles, in an unbound state or as an aggregate or as an agglomerate and where, for 50% or more of the particles in the number size distribution, one or more external dimensions is in the size range 1 nm–100 nm. In specific cases and where warranted by concerns for the environment, health, safety, or competitiveness, the number size distribution threshold of 50% may be replaced by a threshold between 1 and 50%.*

As nanomaterials have fundamentally different properties than other materials, one must be careful when handling them as the risks associated with exposure to them are often unknown. At the end of 2017, the WHO (World Health Organization) published its guidelines on protecting workers from manufactured nanomaterials. We will come back to this in Chapter 6 where we look at the health implications of nanotechnology.

A material that has one dimension in the nanoscale is a film or layer and is essentially 2-dimensional. If it has two dimensions in the nanoscale, then it is a nanowire or nanotube, which is 1-dimensional. If it has all three dimensions in the nanoscale, then it is a particle, or nanoparticle, which we describe as being 0-dimensional. We will see examples of all three types of nanomaterial, why they are useful and where they are to be found. When the nano-hype started in the mid-1990s, there were a number of products on the market that purported to have nanothings in them to improve their performance, but in many cases, the *nano* was really *micro*, in which case the so-called improvement was marginal and did not justify the additional premium. When concerns about the safety of nanomaterials started being raised in the mid-2000s, those products stopped using the word *nano* in their marketing materials. As of the 11th of July 2013, any cosmetics having ingredients in nanomaterial form must indicate so with "nano" on the label. As most cosmetic products contain nanoparticles in some form or another, this regulation has had a profound impact on the industry. Coupled with the EU ruling from the same time, all claims made by manufacturers regarding the efficacy of their products must be substantiated. It is in fact rather shocking to think that until then, it was not necessary to do so.

There will always be individuals with fertile imaginations, and when they are scientists, this can lead to hype that non-scientists assume is correct, as boffins are always right, aren't they? When they are non-scientists, it can lead to ill-informed hearsay, or if common sense is applied, nothing more than a ripping good read. An example of a harmful scenario was in Eric Drexler's book "Engines of creation" from 1987, which is an inspiring read that contains a good deal of superb science, but which also does raise questions that we now know are not actual possibilities. In it, he proposes the concept of molecular machines, and other tiny machines that are capable of building copies of themselves atom by atom and uses molecular modeling/drawing tools to make molecules that look like familiar machines. The exposition of the new field of nanotechnology is indeed very interesting, and the author's take on how technologies become adopted is insightful. However, the material which has some pseudo-scientific components to it does lend itself to be misinterpreted, so it was, and it was seen as saying that nanotechnology was a cure-all that would revolutionize the world around us, but in the wrong hands could lead to world destruction.

To be fair, most scientists are cautious about making such claims, and all Drexler was doing was to look at the possibilities. The key error here, which was a collective one, was the lack of a scale of likelihood of any of the supposed things actually happening. Drexler's hypothesis was based on the assumption that simply by placing atoms in the right positions, they would join together, a bit like bricks (Figure 4). As was pointed out in the public and frank exchange of views between Drexler and Richard Smalley (Nobel Prize winning Chemist who discovered C_{60}), this completely ignores Chemistry — atoms don't just form bonds if you push them together — there has to be a reaction, and not all atoms react together.

Figure 4.   One of Drexler's proposed molecular machines — a planetary gear mechanism. The individual colored balls represent different atoms. This is a drawing of a hypothetical molecule — no such thing actually exists.

Nonetheless, these images of molecular machines persist to this day and in the right context can be enormously useful guides. The 2016 Nobel Prize in Chemistry was awarded to three chemists for their work on the design and synthesis of molecular machines. Jean-Pierre Sauvage, J. Fraser Stoddart, and Bernard L. Feringa worked independently in Europe and the US to develop molecular motors which they demonstrated as early as the 1990s. These were the first steps toward Feynman and Drexler's vision of molecular machines, and together with the STM we can visualize these machines in action. However, the sort of machine envisaged on the previous page is a long way off to say the least, and as mentioned earlier, it is not even clear that it will ever be possible to create such a thing.

One rather facetious way in which molecular machines is being used is the so-called "Nanocar race", which had its first ever *event* in April 2017. Four teams from around the world designed and synthesized organic molecules that look like cars (sort of), and then used a scanning tunneling microscope to push them around on a surface. The first to cross the finish line won! The images are all copyrighted, so it is worth having a look at the website, (nanocar-race. cnrs.fr).

On the topic of daft apocalyptic claims, the analogy that jumps to my mind is that of the end of the world as "predicted" by the Mayans. It had been suggested that their so-called long-count calendar would end on the 21st of December 2012, and this suggestion became part of popular culture, forming the plot of multiple movies, books, and TV series, including the X files. When we look at where this "prediction" comes from, it soon becomes apparent that unsurprisingly it is complete hokum. The Mayan calendar does not end on that day, it simply comes from old legends that the world had been re-created three times since the beginning of time, and that this happened on a 5,126 year long cycle. There was never any suggestion that this would happen again, nor any real belief that it had ever happened anyway. Unfortunately, it makes a great story, and there are those who like a bit of intrigue and who are only too happy to perpetuate such stories, which then become "common knowledge". It is safe to say that all scholars who have looked into this have unanimously agreed that there is absolutely nothing in it apart from a misinterpretation of what the Mayans believed in. I guess it was rather inconvenient and left a number of people feeling a bit foolish when the world didn't just behave and end in 2012.

Why did people misinterpret Drexler's ideas? When a scientist draws a molecule that looks like a machine and then says such things can be made someday, one tends to assume they must know what they are talking about. Specifically, in relation to this book, an example of the non-scientist hype route was Michael Crichton's "Prey". In that book, some misguided scientists create self-replicating nanomachines which eventually destroy all materials in their

(a)   (b)

Figure 5.   (a) Michael Crichton's prey — about self-replicating nanomachines (gray goo) taking over the world. A ripping great yarn, but with absolutely no basis on reality! (b) As the half-life of DNA is not particularly long, Jurassic Park won't be opening anytime soon.

path, including all life on the planet. A great story, *but with absolutely no basis in reality*. The only structures we know that can self-replicate given the correct conditions are DNA and viruses — we simply do not have the technology to create anything that matches them, *nor do we want to*. Bear in mind that the same author wrote "Jurassic park" nearly 25 years ago, and there are still those who believe that some rich, evil megalomaniac is on the verge of resurrecting dinosaurs (Figure 5).

Genetic research has shown us that DNA simply doesn't last that long, so even with the impressive recent skeleton finds (for example, the T-Rex called Jane found in 2001, the most complete specimen ever found), there is *no possibility* that we will ever find DNA that is intact enough to do anything useful with. The specific measure of this is the *half-life* of DNA, which is the time taken for half the bonds holding the strands together to break, which is currently approximately considered to be 521 years, assuming a temperature of 13°C — this number can be extended to hundreds of thousands of years under freezing conditions, and if encased in amber and frozen, possibly even longer. Still, it is difficult to imagine conditions on earth where enough dinosaur DNA would last the 65 million years it has been since the last extinction level event.

We have now said quite a few times that nanosized pieces of many materials behave differently to larger-sized pieces of the same material. If we can gain an understanding of *why* this is true, then we can have some predictive power over what to expect when we create a nanomaterial. We have addressed this in the earlier chapters, but to recap, there are two primary drivers — surface area and quantum effects. To see what I mean about surface area, consider taking a block of material as shown below. It is cubic, with a side of 1 m, so has a volume of 1 m^3. Each face has an area of 1 m^2, so the overall surface area is 6 m^2 (Figure 6).

If we now split this up into little cubes of side 10 cm, we retain the same volume, but end up with significantly more surface area, as illustrated in Figure 7.

Each of the cubes has a volume of 10 cm $\times$ 10 cm $\times$ 10 cm = 0.1 m $\times$ 0.1 m $\times$ 0.1 m, which is a volume of 1000 cm^3 or 10^{-3} m^3. There are a thousand such mini-cubes, so the overall volume is the same as before, at 1 m^3. However, if we

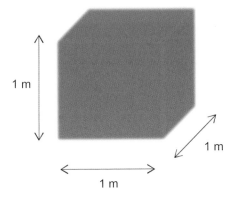

1 m

1 m

1 m

Figure 6.   1 m^3: Volume is 1 m^3 Surface area is 6 m^2.

The same 1 m3

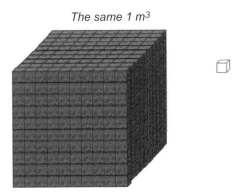

Figure 7.   The same cube is now split into 1000 (10 cm $\times$ 10 cm $\times$ 10 cm) cubes. The combined surface area is 10 times bigger than the 1 m cube on its own.

now look at the surface area, there is a big difference. Each cube has a surface area of $6 \times (0.1 \text{ m} \times 0.1 \text{ m}) = 0.06 \text{ m}^2$. There are 1,000 mini-cubes, so the overall surface area becomes $1000 \times 0.06 \text{ m}^2$, which is 60 m^2, an increase by a factor of 10.

If we were to have made the cubes 1 cm on a side instead of 10 cm, then there would be 1,000,000 of them inside the 1 m cube, each with a surface area of $6 \times (0.01 \text{ m} \times 0.01 \text{ m}) = 0.0006 \text{ m}^2$. This yields a total surface area now of 600 m^2, a further increase by a factor of 10. So, every time we split a structure up into subunits that are 1/10 of its size (on a side), we increase the overall surface area by a factor of 10.

Imagine now if we were to keep on doing that and now split our 1 m cube into units that are just 10 nm on a side — their combined surface area would be 100,000,000 times larger than the original cube! Who cares? Well, surfaces (The Nobel Prize winning Physicist Wolfgang Pauli once said that *God made the bulk but surfaces were made by the devil*) tend to have different properties than bulk materials due to the fact that the atoms on a surface are not surrounded on all sides by other atoms, so they tend to be more reactive. This is the reason why catalysts tend to be in the form of nanoparticles (Figure 8). Re-examining this idea, now imagine a cube of side L. The volume is L^3 and the surface area is $6L^2$. The ratio of surface area to volume is therefore $6/L$. As the cube is made smaller and smaller, this ratio will get larger and larger and by the time L has dimensions in nanometers, the surface area will be dominant over the volume. By the time we get to lower-dimensional structures such as carbon nanotubes and graphene (which is a sheet of carbon atoms that is one atom thick), they are all surface — hence their exceptional properties.

An example of this is well known to all those who take sugar in their tea or coffee — a habit that I was broken out of what with being married to a dentist.

Figure 8.   Inside a catalytic converter from a car. This is structured to maximize its surface area and is coated with the catalyst material in nanoparticle form.

Figure 9.   Powdered sugar dissolves much faster than sugar cubes due to *their having* a much greater surface area.

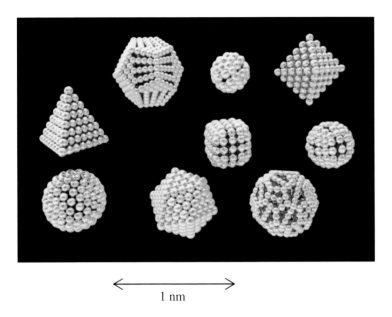

1 nm

Figure 10.   In particles less than around 5 nm across, the individual atoms start to become notice-able. Shown here are gold nanoparticles — each ball is a single atom, and which shape a given col-lection of atoms will take depends on the number of atoms in it.

Experience tells us that powdered sugar dissolves much faster than sugar cubes, and the reason is as described above — the powdered sugar has a much higher surface area, and therefore melts far faster than the cube (Figure 9).

By the time nanoparticles are only a few nm in size, the individual atoms start to become noticeable. This is illustrated in Figure 10 that shows the atomic arrangement in particles around 1 nm across, where all of the atoms on the surface are surrounded by atoms beside them and between them and the center

(a)   (b)

Figure 11.   (a) A painting of the Houses of Parliament in London by Claude Monet from around 1900 — London was notorious for its "pea soup" smog from coal burning; (b) A view of Shanghai early in the morning showing smog.

of the particle but have no atoms between them and the outside. We say that the atoms on the inside are *fully coordinated*, and have bonds to several other atoms, whereas the surface atoms are not fully coordinated, and are not as stable. As a consequence of this, the atoms on surfaces tend to be more reactive than those in the bulk. In some materials, the atoms compensate for this by moving slightly closer together (e.g. Gold) or by pairing up and forming "dimers" (e.g. Silicon).

Nanoparticles are to be found in very many places for a variety of reasons. For a start, a large number are to be found in the air we breathe, and those particles come from a number of sources, mostly man-made. An exposition of the types of particulates in the air are to be seen in Figures 11–13. Particularly high levels of smoke and fog combine to form "smog", which has historically been a major health concern in developing countries of densely populated areas with large numbers of factories or motor vehicles, from Victorian London to Los Angeles, San Francisco, Mexico, and Beijing to mention a few. In warm dry areas of bright sunshine, matters are compounded as the emissions (particularly Nitrogen Oxides) in the atmosphere are activated by sunlight to react further, making more nanoparticles and creating what is known as a "photochemical smog" — particularly prevalent in California — one of the reasons why, despite my having been in San Francisco three times now, I have never seen all of the Golden Gate bridge — it's always been draped by a brownish haze. Improvements in fuels have led to an overall reduction of these problems in recent years, but they still constitute a genuine health concern, for good reasons.

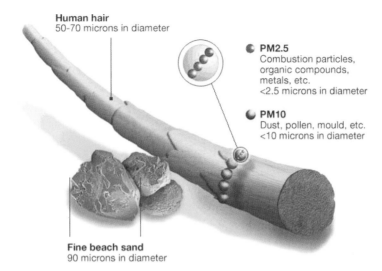

Figure 12.   Particulates in air are placed into two categories based on their size, depending on whether they are bigger or smaller than 2.5 microns in diameter. Compared to the diameter of a strand of hair, they are all very small! (*Image Source*: US EPA).

Petrol-burning engines have become more efficient and less polluting in recent years, driven by government policy in the developing world. Diesel engines unfortunately are not so squeaky clean and produce roughly 20 times more nitrogen oxide compounds than petrol engines. The concern regarding fuels is two-fold — highly reactive nanoparticles and volatile chemicals. The size range of the harmful soot nanoparticles in engine emissions is of the order 5–200 nm, so is right within the range we are interested in in this book. The emissions from vehicles are categorized as particulates and volatile compounds. The particulates are split into two size categories: PM10 and PM2.5 which refers to particles between 2.5–10 microns in diameter and 0–2.5 microns in diameter, respectively (Figure 12). The key issue with the smallest (PM2.5) particles is that they are small enough that once they get into the lungs, they can enter the bloodstream, and from there can go anywhere in the body and start to cause trouble. We will see later when we look at nanomedicine that small particles can trigger the formation of cancer cells — let's not forget that one of the biggest killers is lung cancer which to a large extent is caused by particulates in smoke. Not all nanoparticles are bad of course, and that efforts are underway to artificially create ones that can actually cure cancer — we will explore this in Chapter 7.

In Figure 13, I have presented a who's who of airborne particles, where the three columns on the left are nanosized, and unfortunately, these are the most harmful ones. The worst offenders are not nanoparticles in the true sense but are the toxic gases/fumes emitted by engines, primarily compounds of nitrogen and

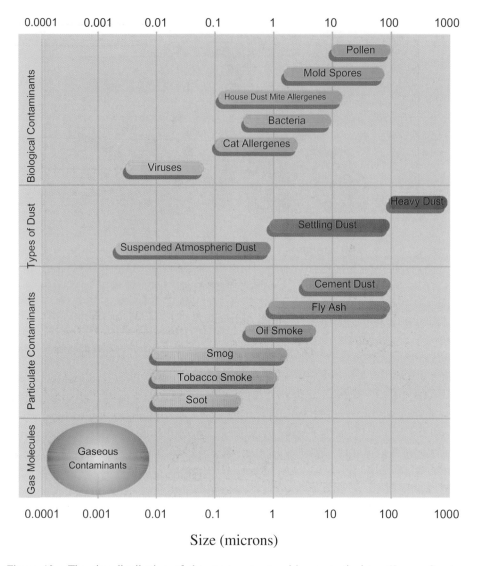

Figure 13. The size distribution of the most common airborne particulates (*Image Courtesy*: GFDL, https://en.wikipedia.org/w/index.php?curid=48987967).

oxygen. The UK air-monitoring network (called "UK-AIR") measures the nano-particle and particle count as well as the specific particulates and contaminants in the air right across the country, and the data is freely available.

In Figure 14, there is some data showing the typical particle count to be found over a four-month period in Bloomsbury in London. As you might expect, the count is highest when outside, due to traffic. Most of the nanoparticles in the atmosphere are carbon-based and are the result of internal combustion — i.e. they are soot produced by engines and have a typical size in the range 20–30 nm.

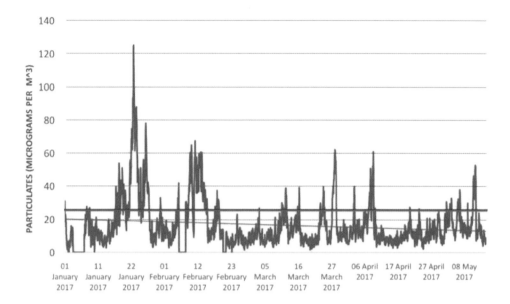

Figure 14.   Measured PM2.5 particulates in the air in Marylebone, London, from the period January–May 2017. The EU suggested maximum average mass is 25 mg/m³, as indicated by the red line.

Note: Airborne particle count recorded during a 4-month period in London Bloomsbury. Data from UK-AIR (DEFRA).

Over this period, the particle count went as high as 125 mg/m³, as compared to the average value of around 15 mg/m³. Should we be concerned? We will look at this later when we explore nanoparticles, but the mounting evidence is that yes, we should, but the bigger problems are from toxic gases such as nitrogen-oxide compounds rather than nanoparticles. Nonetheless, there are recommended "safe" levels for exposure to such particles. In the EU, this is 25 mg/m³. In the US, it varies by state, but is typically in the range 15–35 mg/m³. It should be borne in mind that as with almost everything, the notion of a "safe" level is somewhat arbitrary and is subject to constant changes. It is known that a rise of 5 mg/m³ is associated with a 7% increase in mortality (*source*: NHS UK), so there is a continuous drive to improve air quality. Although these particles are produced by combustion, it is not just cars that are to blame — we must also consider that every home that has central heating is also burning fuel, creating these particles and contributing to global pollution.

We will come back to nanoparticles later, but first I would like to deal with the other forms of nanomaterials — films and nanowires/nanorods.

Thin Films —The First Step Toward Nano

Films are thin layers of materials, that can either be free-standing or in the form of a coating. A film is shown in Figure 15 and in order to be considered as a nano-material, it needs to be less than 1000 nm and usually less than 100 nm thick.

< 100 nm

Figure 15.   A thin film nanomaterial.

Examples of films of materials are commonplace, the most familiar are cling film and aluminium foil, and you could argue that paper is a film, as it is much thinner than it is wide. Cling film is made from LDPE, or low-density polyethylene, and is typically 12.5 microns thick, and aluminium foil is typically 24 microns thick (Figure 16). By contrast, typical paper that we use in a printer is around 100 microns thick, and newspaper is around 60 microns thick. When we start getting thinner than around 20 microns, many materials become very weak and can't support their own weight (cling film is a very good example of that). That does not mean they are not useful though, as many coatings have thicknesses in the range of a few tens of microns or below. An example is the silver plate used on cutlery — the thickness is typically in the range 10–30 microns depending on the quality. Another example is paint — spray paint has a thickness of the order 25 microns. This is all illustrated in Figure 17, putting things into relative context.

Generally speaking, if the purpose of the film or coating is aesthetic, then much thinner films will be used to reduce costs. A primary example of this is the decorative chrome plating as used in the automotive industry and in many household cases, e.g. radiators and towel rails. In these cases, the film thickness is usually in the range 50–500 nm, so falls within the definition of a nanomaterial.

Do films with thickness less than around 1000 nm have different properties than thicker films? The answer is that some properties are different and others are not. The function of thin metal films is generally to provide a shiny surface. They do this very well, as the optical properties of these films do not particularly depend on their thickness until they become even thinner, of the order 10 nm. What I am getting at therefore, is that although we commonly encounter coatings that are thin enough to be classified as nanomaterials, this is not always important as their relevant properties have not been modified by virtue of being thin. Other properties will change though, such as their electrical resistance or conductance (1/resistance), as shown in Figure 18. Mechanical strength is also very sensitive to thickness and size.

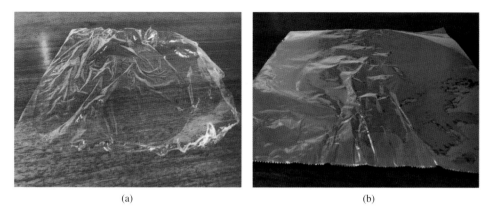

(a)   (b)

Figure 16.   (a) Cling film — an everyday thin film, usually around 12.5 microns thick; (b) Aluminium foil — another everyday thin film, usually around 24 microns thick.

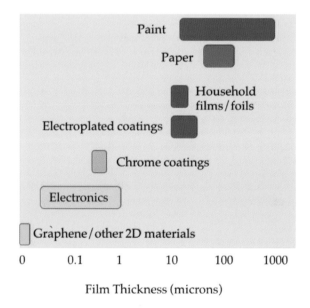

Film Thickness (microns)

Figure 17.   Everyday items that can be considered to be films, and their thickness range.

The electrical resistance of a given object generally depends on its geometry, size, and the material it is made out of. Specifically, the resistance increases as the length of an object increases and its cross-sectional area decreases. All materials have a property known as *electrical conductivity*, which as the name suggests, is a measure of how easily electricity can flow (conduct) through them. It had been thought until the 1960s that electrical conductivity was a constant for a given material, and therefore, once you knew the geometry and

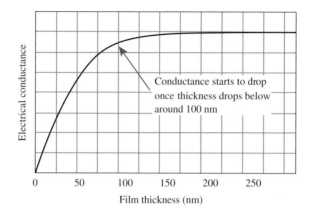

Figure 18. The variation of electrical conductance (1/resistance) with film thickness for a thin film. The electrical properties of metals deteriorate as they approach nanometer dimensions.

size of an object, the resistance could easily be predicted. This is true of objects above a certain size, which happens to be around 100 nm — dictated by the mean free path of the electrons. Below this, as a film of material is made thinner, the electrical conductivity starts to drop rapidly. This has implications for the semiconductor industry that we will explore shortly. In the most extreme case, if a conductor is made small enough (less than around 1 nm), it will become insulating.

One of the reasons for this decrease in conductivity comes from the fact that thin films of metals are grainy. As we saw earlier, this granularity has a substantial influence on mechanical properties also. The tiny gaps (much less than 1 nm) between grains are called grain boundaries as illustrated in Figure 19, and they have a high electrical resistance.

Generally speaking, thin films of metals are produced by some sort of vacuum sputtering process, where there is a target of the metal of interest, which is bombarded by an ion beam (usually Argon, as it is inert). This causes atoms to be ejected from the target, and to be deposited on an appropriately placed object (Figure 20). If we were to observe the deposition on that object, it usually follows these steps:

- First sputtered atoms land on object and stick to it/become embedded in it.
- More sputtered atoms arrive and do the same, some stick more to the other sputtered atoms and start to form little *clusters* where the atoms arrange themselves in rows.
- Clusters act as nucleation sites, and attract incoming sputtered atoms.
- Clusters start to grow and form grains.
- Grains eventually merge, forming a continuous film and large ones subsume smaller ones — this becomes more noticeable as the film gets thicker.

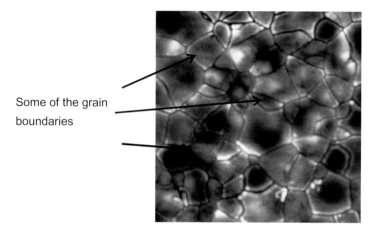

Some of the grain
boundaries

Figure 19.   Grains in a film. The more grain boundaries there are in a film, the higher is its electrical resistance.

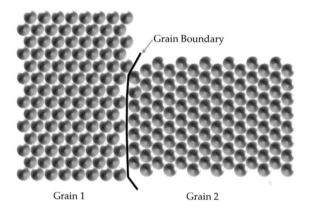

Figure 20.   A grain boundary at the atomic scale. The circles represent atoms. Within each grain, the atoms are arranged in regular positions (crystal).

The only difference between grains in the same film is that the atoms are arranged along different directions from one grain to another. As more material is sputtered, the grains grow, and they start to encounter each other. They don't usually join up nicely due to those different directions of the atomic rows, so often a little gap is left between grains, which we call a grain boundary. If the film is given enough energy, e.g. by heating, then the atoms can cross the grain boundary and whichever grain the atoms can move fastest from will tend to grow the most, therefore subsuming the other grains.

Due to this way in which films grow, it turns out that there is an almost 1:1 relationship between the average size (width) of grains and their thickness. In other words, the thicker a film is, the larger the grains get in *all* directions, therefore reducing the number of grain boundaries (Figure 21).

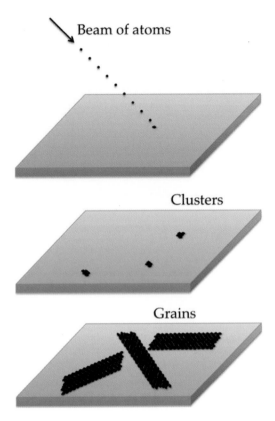

Figure 21.   How films grow and become grainy. During deposition, a beam of atoms is directed at a target surface (here shown in blue), where they tend to gather together forming little clusters which grow into individual grains. There is often a 1:1 correlation between the thickness of the film and the lateral size of the grains.

This is observed as an increase in electrical conductivity as a film gets thicker, and it reaches a steady value once the thickness is above a hundred nm or so, or 2–3 times the mean free path of the electrons. This causes issues when we want to make films as thin as possible, for example, in microprocessors where we want to pack as many components as possible into a small space. As we make films and wires thinner and narrower, their electrical resistance increases significantly. This leads to more heating and power losses, and hence a drop in efficiency. This is something we will look at later in the book when we consider computers from a nano-perspective, but the bottom line is that nano is not so good when it comes to electrical properties.

Why Are Nanostructures so Strong?

Thin films of materials have desirable mechanical properties when compared to thicker films of the same materials. This is shown in Figure 22 which plots

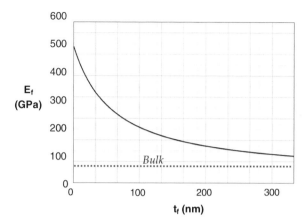

Figure 22.   The strength (Young's modulus) of a gold film versus its thickness, compared to that of a bulk piece of gold. The mechanical properties of metals massively improve as they approach nanometer dimensions.

mechanical strength versus film thickness for gold films. Similar characteristics are seen for all metals. The standard measure of strength of a material is a quantity known as the Young's modulus, otherwise known as the elastic modulus, given the symbol E, which is defined as the ratio of stress-to-strain. Stress is the pressure applied to a material and strain is the proportional change in its length as a result of this pressure. This is measured as the change in length divided by the original length. A stiff material will deform less than a soft material when a force or pressure is applied to it. The units of Young's modulus are the same as those of pressure — the Pascal. As an example, steel has a Young's modulus of around 200 Gigapascals, that's a 2 followed by 11 zeros, written as 200 GPa. This means that if we hang a force equivalent to 50 kg from a steel rod 2 mm in diameter and 1 m long, it will extend by around 0.8 mm, which is barely noticeable. However, if we use a nylon fiber with the same dimensions, the deformation will be more like 8 cm, as nylon has a Young's modulus around 100 times lower than steel. Of course, there is a chance that the nylon will break before it stretches that much as there is a limit to how much deformation or strain any material can withstand. Once the elastic limit of a material is reached, it will stretch irreversibly and eventually fracture. In the examples above, the stress applied was a 50 kg force acting over an area of 3.14×10^{-6} m^2, which is a pressure of 159 MPa, which is actually rather low, and in fact the steel could withstand a higher force, as the stress at which it breaks (its yield stress) is around 450 MPa. Nylon unfortunately has a rather low yield strength of around 45 MPa, so in fact it will not extend by 8 cm — it will break long before that.

What nanotechnology has to offer is that we can significantly enhance the properties of an existing material by adding nanomaterials, and without too much additional cost. This is the principle behind alloying — mix two or more materials together to get the best of all. This is also true when forming composite materials. Going back to our thin film of gold, what we can see from Figure 22 is that a 200 nm thick film is twice as stiff as a bulk film. Decreasing the film thickness to around 100 nm doubles the strength again. The properties keep improving as thickness reduces until we reach a thickness of a few nm, below which point the film becomes very weak. There are a number of reasons for the enhanced strength of thin films, but the single most important one is rather mundane — as grains grow and films get thicker, more and more defects and imperfections start to arise so thicker films are less perfect.

It is these imperfections that give rise to weaknesses in real materials. The effect of such imperfections was dramatically seen on April the 14th, 1912, when the largest iron structure ever built met its Waterloo. I am of course referring to the *Titanic* (Figure 23). It was constructed using wrought iron and steel, and as the builders (Harland and Wolff) were charged with building three enormous ships at essentially the same time, there was a lot of pressure on their steel supplies, so it appears that less than optimal grade steel and iron was used.

The particular steel and iron that was used in the construction of the *Titanic* was unusually high in sulphur content, from the slag that had not been properly removed from the iron ore during smelting. Slag is the collective name for the most common impurities in iron ore, and consists of oxides and sulfides, and is a glassy, brittle substance. Studies carried out on some rivets that were brought up from the wreck in the early 1990s demonstrated that the slag was in the form of large particles, and the rivets shattered rather than tore, so they were brittle. This was partly because of the low temperature of the sea (around −2°C) causing

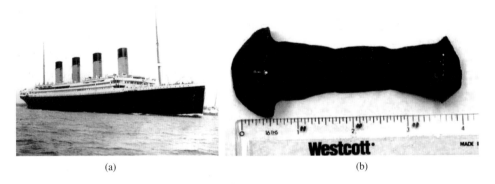

(a)   (b)

Figure 23.   (a) The Titanic as it left Southampton on the 11th of April, 1912; (b) A hull rivet from the Titanic, recovered from the wreckage (Image from InFocus magazine, 2007).

the slag particles to become even more brittle than they usually are, rendering the entire rivets weak. Inconsistencies in the manufacture of the rivets meant that there was enormous variability between different batches, and that in some cases the slag content was up to three times higher than usual. We can describe the rivets as being a composite material — soft iron with inclusions of hard slag — an example of when a material is made worse by the addition of another material.

This was compounded by the fact that the ship builders were so pressed for time that they hired a large number of either unskilled or insufficiently skilled workers to make the ship. It appears that the puddlers (the people who made the rivets on-site) were inexperienced and overworked, and so cut corners — the rivets were formed at too low a temperature, and the molten iron was not stirred sufficiently to separate out the slag, resulting in large slag particles embedded throughout the rivets. This rendered them weak and probably at almost breaking point when they were installed, never mind when the iceberg was struck. It has been suggested that had appropriate rivets been used, not as many of them would have popped, and the *Titanic* would probably have been able to stay afloat for long enough for the nearest ship, the *Californian,* or the first ship to arrive, the *Carpathia*, to reach her on time to rescue all on board.

Therefore, as a film gets thinner, it tends to have more grain boundaries, but otherwise fewer impurities or defects in its structure. The latter outweighs the former, so the strength of a film increases even though its conductivity decreases.

Keeping Things Clean

Nanometer-scale thin films are in common use, mostly in electronic components, but they are also to be found in unexpected places. Common everyday examples are self-cleaning glass and anti-reflection coatings — on glasses (spectacles), computer/tablet/phone displays, car windscreens, and even windows. In the case of self-cleaning glass, the surface of the glass has been modified to render it *hydrophilic*, which means "water loving". What does this mean in reality? Well, when water lands on a hydrophilic surface, it spreads out, or *wets* the surface, and can then sheet off it. By contrast, a *hydrophobic* surface is one which water doesn't like to sit on, so when water is placed on such a surface, it forms droplets (Figure 24).

One of the most commonly quoted examples of a hydrophobic surface is the lotus plant. The surface of the leaves has a coating with micron and nanometer-scale features and is so rough that water cannot form a nice layer on it due to surface tension — it is energetically more favorable to form droplets. This is commonly called *the lotus effect*. As these droplets roll off the leaf, they pick up debris and carry it away, thus cleaning the leaf.

(a)   (b)

Figure 24.   (a) An example of a hydrophobic surface: The Lotus effect — the coating on lotus leaves is so rough that water prefers to form droplets on it, which can roll off and carry away dirt; (b) A hydrophilic surface — the pitcher plant. This contains grooves aligned in such a way that they channel water downwards into the sac. When water lands on this plant, it wets the surface — i.e. it spreads out in a thin layer.

Another example of nature being clever is seen in the pitcher plant, where the surface is covered with nanometer-scale grooves that are aligned vertically, and their net effect is that water spreads easily along these grooves, rendering the surface *superhydrophilic*. As a result, insects slip down it to their doom and they can't rely on any capillary force to help them. The Darkling beetle from the Namib desert has gone one step further and uses both effects. It has micron-scale bumps all over its surface which are hydrophobic, but interestingly, cause water vapor from the air to condense as droplets. It also has nanometer grooves along the length of its body which then create a hydrophilic pathway for those droplets to flow into the mouth. There are efforts to try to mimic this in order to harvest water vapor from the air in desert areas where drinking water is scarce.

Therefore, there are arguments for wanting a surface to be either hydrophilic or hydrophobic, as both can result in a clean surface. We just don't want something in between, as in that case, water will stay there, stuck to dirt on the surface. As rainwater evaporates, it leaves even more dirt behind, as there is always dust at the center of raindrops. The energy required for water vapor to spontaneously condense into droplets is enormous, but if it has something to condense on, or *nucleate*, then it happens rather readily as we all know. Hence, the first step toward keeping glass surfaces clean is to prevent water from staying on them and eventually evaporating.

This principle was taken and extended back in the mid-1990s by the glass manufacturer Pilkington, who created the first self-cleaning glass, which they called *Activ*™. This had a 20–30 nm of nanocrystalline titanium dioxide, TiO_2 on the glass surface (Figure 25). In bulk form, this is a white powder, but when thin enough (below around 50 nm), it is transparent. It also helps that it is non-toxic, hence its use in many cosmetics and food colorings. We will see in Chapter 7 that

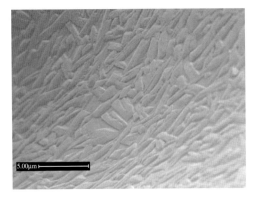

Figure 25.   The titanium dioxide coating on self-cleaning glass. This coating makes the surface hydrophilic, so water wets the surface and can sheet off it easily.

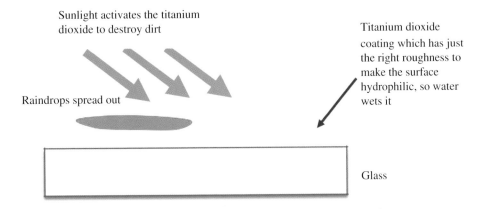

Figure 26.   The Principle of self-cleaning glass. The roughness renders it hydrophilic so water spreads out on the surface and can sheet off it quickly without leaving streaks, and the titanium dioxide coating is activated by sunlight to break down organic material by oxidizing it.

this statement is not without controversy. The titanium dioxide is sputtered on glass and forms a layer that is hydrophilic, with the added bonus that when sun shines on it, it *photocatalyses* or chemically breaks down dirt (Figure 26).

The Barlow shed at St. Pancras railway station in London is a fine example of the implementation of self-cleaning glass, and this was installed between 2001 and 2007 (Figure 27). Bearing in mind that in order for this kind of glass to be effective at cleaning, it must have water falling on it, so vertical structures such as the Gherkin building are not suitable. The Barlow shed is curved, so when it rains, water will naturally run down it and spend long enough on it to pick up any dirt particles, but not so long that it evaporates and leaves dust behind.

From a domestic perspective, self-cleaning glass appears to be a good choice for slanted windows, such as those found in skylights or conservatory roofs, but

Figure 27.   The Barlow shed at St. Pancras train station in London — this is where the Eurostar leaves from. The roof windows are made using self-cleaning glass.

there is little evidence that they make sense anywhere else in our homes, apart perhaps from ground-floor windows. The cost of modifying glass to make it self-cleaning is coming down all the time, and currently adds a premium of around 10–15% to the price, so soon this may well become commonplace (Figure 27). The coating technology has improved so much in the past 3–5 years that now the films only need to be around 15 nm thick and can still be equally as effective as before. As with all coatings though, it can be damaged and scratched. However, with careful handling, this is not a problem. Recent work suggests that this coating also has photo-active anti-bacterial and anti-viral properties, and while the jury is out on that one, it does offer the intriguing possibility that we could use coatings on our windows to help protect us from airborne pathogens (Figure 28).

Seeing Things More Clearly — Anti-Reflection Coatings

The second everyday application for nanometer thin films on glass is anti-reflection coatings. The optical coatings market is worth several billions of dollars per year, and of that, anti-reflection coatings comprise the largest part. These are thin films of transparent materials that are coated on glasses, camera lenses, telescope and microscope lenses, laptop and tablet screens, and on many optical machines. Their function is to remove unwanted reflections or glare, and their operation is based on wave interference.

When light is shone on any surface, some of it will be reflected. When talking about the optical properties of materials, the quantity known as *refractive index* is important. It is literally a measure of how much light bends when it passes from one

Figure 28. Self-cleaning glass in action on a sunny day after a rain shower. The pane on the left is normal glass with water stuck to dirt particles, and the pane on the right is self-cleaning glass where the dirt particles have been photocatalytically broken down by sunlight activating the titanium dioxide. The by-products of this process have been washed away by the rainwater as it sheeted down the window.

material to another. The larger the refractive index of the material the surface is made from, the more light will be reflected from it. Glass has a refractive index of around 1.5, and around 4% of incident light is reflected from it, although this can approach 100% if we are looking at a grazing angle. Metals have a refractive index of several thousands, so they reflect the majority of the light incident on them.

The particular problem with glasses is light reflected from the part of the lens closest to us, as it means we see glare. A thin coating, usually an oxide material such as silica or magnesium oxide, is sputtered or chemically sprayed onto the lens, to a thickness of around 100–150 nm. The idea, as shown in Figure 29, is that light now has two surfaces to get reflected from: the surface of the coating and the surface between the coating and the glass. Each of these materials has a different refractive index, so some light will be reflected from both of those surfaces. The thickness of the coating is specifically chosen to be ¼ of the wavelength of light, so the light that gets reflected from the glass will have traveled approximately ½ wavelength more than the light that was already reflected from the coating. This leads to destructive interference, as the peaks in the latter light wave will now coincide with the troughs from the former one. In other words, the light rays reflected from those two surfaces will cancel each other out, which we observe as a lack of reflected light.

These coatings serve two purposes — they reduce glare for us, and for people looking at us. These coatings are often capped with a thin layer of a fluorocarbon material which is oleophobic — *oil hating*, to reduce the appearance of

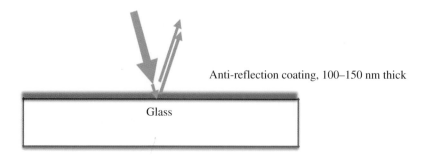

Anti-reflection coating, 100–150 nm thick

Glass

Figure 29.   Principle of Anti-Reflection Coating: Some light reflects off the surface of the coating, and some off the top surface of the glass, and these light rays cancel each other out by interference — an effect due to the wave-nature of light.

fingerprints. Such coatings are widely used in touchscreen devices such as smart-phones and tablets, to enable easy cleaning. This is really just an extension of the Teflon® coating used in cooking utensils, which is PTFE, or polytetrafluoroethylene, but in a much thinner format. The PTFE coating we are familiar with is typically around 0.1 mm thick, whereas the coating on smartphones and glasses is just a few nanometers thick.

The cost of these layers is considered to be relatively expensive, and for a typical set of glasses/lenses which cost around £100, the coatings cost £10–30. This is more to do with the fact that the machinery used to produce these coatings, which is carried out using a vacuum deposition process, can only take a small number (a few hundred) of lenses at a time, and lenses tend to be made to order. Many other industrial processes, such as formation of the gorilla glass used in the iPhone, apply the coating to a large number of pieces simultaneously, and they are all the same shape and size, so the process can be scaled up more easily, and therefore made cheaper.

Ultimate Thin Films — Just One Atom Thick: Graphene

The ultimate thin film, that has received an enormous amount of media attention and funding, is of course graphene. It has been touted as the strongest material ever discovered and has electrical properties far superior to those of silicon. Graphene is the name we give to the individual sheets of regularly arranged carbon atoms that collectively make graphite. These sheets are just one atom thick, and the bonds between the carbon atoms are even stronger (shorter) than those in diamond. The atoms are arranged in a chicken-wire pattern, as shown in Figure 30. The key with graphene is that as the bonds are so strong, it is incredibly strong under tension. We have been experimenting with graphite for

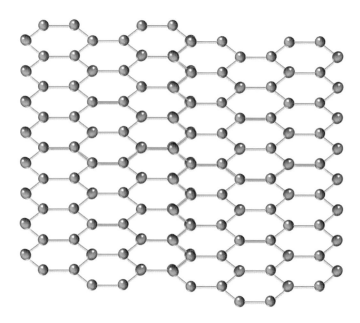

Figure 30.   The arrangement of carbon atoms in graphene is similar to the pattern in chicken-wire fencing.

many years, and in fact it is often used as a support for samples that we wish to image using scanning probe microscopes, as we will see in the next chapter.

The story of graphene is that it had been shown over 80 years ago that no truly two-dimensional material could ever exist, as thermodynamics forbids it. It was assumed for many years that this would make any single-atom thick material too unstable to ever be made, as it should either melt or split up into fragments. Nonetheless, as graphite is composed of single-atom thick layers that are weakly coupled to each other, scientists have been calculating the properties of those layers since the 1940s in order to better understand the electrical properties of graphite. As early as the mid-1970s, single-layer graphene sheets were being grown on other surfaces and studied by a range of techniques.

However, when graphene (or indeed any 2D layer) is sitting on top of a substrate, it is not truly 2D, as it is part of a 3D structure. The assertion about graphene was that a free-standing layer could not exist. However, this was proved wrong in 2004 by Andre Geim and Konstantin Novoselov, who managed to extract single-atom thick sheets, for which they received the 2010 Nobel Prize in Physics.

How can we explain this apparent contradiction? If thermodynamics (which I can assure you, is never wrong!) says that 2D materials cannot exist, then how come people have made them? Who's wrong, thermodynamics or the scientists making graphene? As it happens, and as you've probably guessed, it's not quite

(a)   (b)

Figure 31.   (a) Two single-layer (1-atom thick) pieces of graphene, as seen using an optical micro-scope. Each piece is just a few microns across; (b) The same two pieces of single-layer graphene as observed using an Atomic Force Microscope (AFM).

that simple. The best way to answer this enigma is to say that graphene isn't *really* 2D. Yes, it is one atom thick, but when it is free-standing, two things happen — (a) it deforms and becomes rippled and (b) the atoms vibrate out-of-plane, sampling that third spatial dimension.

Even when graphene is sitting on certain surfaces, such as silicon dioxide, which it only has a weak interaction with, it can form ripples (Figure 31(a)), which are the white patches shown in the AFM image in Figure 31(b). A point about semantics — strictly speaking, the word *graphene* refers to a single-atomic layer of graphite. There are many cases when this word is misused and used to describe pieces of graphite or multiple layers of graphene. Once it goes beyond a single layer, the properties start evolving toward bulk graphite, so we need to be circum-spect. Almost certainly any commercial product that purports to use graphene is actually using graphene power or granules that are really just tiny pieces of graphite.

How is graphene actually made? There are a multitude of techniques ranging from the decomposition of hydrocarbon gases at high temperatures on metal sur-faces, whereby graphene naturally forms, to the old favorite of mine and the original technique used by Geim — mechanical exfoliation. This is one of those cases of high tech meets low tech, like using chewing gum to plug a leak in a rocket. The idea for mechanical exfoliation goes back some time and many researchers, myself included, have been doing this for years to produce clean graphite surfaces. As we will see in Chapter 5, highly crystalline graphite is one of the most studied surfaces due to the ease with which the single carbon atoms can be observed and the ease with which fresh, clean surfaces can be prepared.

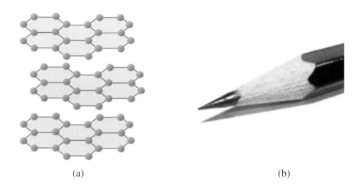

(a)   (b)

Figure 32.   (a) Graphite comprises layers of graphene that are around 1/3 of a nanometer apart, so are only very weakly bound to each other; (b) Pencil lead is made from graphite, as the layers can slide over each other easily and stay behind on paper.

The structure of graphite is shown in Figure 32(a), and it consists of weakly bound graphene sheets. The atoms in the individual layers are held together by very strong *covalent* bonds, whereas the layers are held together by the much weaker *van der Waals* force. This weak binding enables the layers to slide across each other very easily, hence the use of graphite in pencils (Figure 32(b)) and as a solid-state lubricant.

This weak binding is very useful, as we can easily strip layers off to reveal a fresh clean surface. This is traditionally done using adhesive tape, usually either sellotape® or scotch tape®. The procedure is as follows: start with a fresh piece of graphite, place some scotch tape on it, press it on firmly and peel it off, as shown in Figure 33. Graphene is produced by successively peeling the layers left behind on the tape — each time it is peeled, the film halves in thickness, so eventually there will just be one layer left. Imagine just how irritated myself and many others are, as we have been doing the first step for many years, and usually just throw that piece of scotch tape away! The Nobel Prize will have to be for something else I guess.

What is usually done after a few peels is that the layers on the tape are pressed against a piece of silicon, and some graphene can get transferred across that way. This is something that many researchers do for themselves, although a number of small companies, that are university spin-outs, produce and sell these materials.

Size Matters in Sports...

We have seen that graphene is exceptionally strong and is already being used in products on the market. I recently discovered tennis racquets containing graphene in the shaft (Figure 34). The argument is that graphene is the strongest and lightest

(a) A graphite crystal (b) With scotch tape® on top

(c) After peeling the scotch tape® —
some layers come off

(d) After peeling the layer on the scotch tape® twice –the
central piece was the first one, then it was peeled to the
left, then to the right.

Figure 33. Stripping layers of graphite.

(a) (b)

Figure 34. (a) A HEAD graphene tennis racquet. It contains graphene in the shaft to make it lighter
yet stronger; (b) The shaft of a HEAD graphene tennis racquet.

material ever made, so if it is added to the existing material the racquet is made out of, then it can make it lighter and stronger, allowing for an easier swing, and allowing the weight to be better distributed. Does this argument hold up to scrutiny? As with many things, this is a difficult one to answer, and one would assume that there must be some scientific data supporting this claim. At the moment, the "evidence" appears to be that these racquets are endorsed by Maria Sharapova, Andy Murray, and Novak Djokovic. To determine whether they actually make any difference, one would need to carry out automated experiments where the trajectory of a large number of balls struck using such a racquet are measured and compared against a non-graphene containing racquet.

For the last few decades, tennis racquets have been made from carbon fiber reinforced polymer, sometimes called *graphite*, not to be confused with actual graphite. The cost of carbon fiber is around £17 per kilo, and that of graphene typically varies from a few hundred pounds to a few thousand pounds per kilo, depending on whether it is in nanopowder or nanogranule form. This cost is plummeting, and is a figure from early 2014, but given the complexity of graphene, I would expect that it will remain significantly more expensive than carbon fiber. The cheapest graphene tennis racquet I could find was a child's one costing just under £45. The weight of the racquet is 245 g, and from the geometry we can estimate the weight of the shaft — the region containing graphene, to be around 1/10 of that, or approximately 25 g. Given that the pricing of manufactured goods is typically split as 20% materials cost and 80% manufacturing cost, we can estimate that the cost of materials in the racquet is of the order £9, and the shaft accounts for about 90p of that. This means we would expect the cost of the shaft materials to then be of the order 90p for 25 g or £36 per kilo. On that basis alone, we would expect that the shaft contains around 3% graphene.

If we assume that the cost of bulk-purchased graphene is a lot less than the figures above, then it is possible that that figure is closer to 10%, but probably no more. Is that enough to make a difference? Probably not to me, but given how skillful the top players are, they presumably can notice such a small difference. The manufacturers claim that these racquets allow the user to hit the ball with greater precision and more power, and the purpose of the graphene really is just to allow the shaft to be lighter, but without compromising its strength, so to increase the strength-to-weight ratio. Other similar commercial uses include cycling helmets, cycling shoes and skis, although undoubtedly others will follow soon enough (Figure 35).

One thing to bear in mind is that the graphene used in manufacturing is not those well-behaved single-atom thick sheets that we have seen so far but is

(a)

(b)

Figure 35.   (a) Some products containing graphene to increase its strength-to-weight ratio; (b) The sort of graphene powder that is added to materials (mostly polymers) to make them stronger & lighter.

actually a powder containing fragments of what is more like graphite, or is usually several layers thick, so technically should not be called graphene. This graphene nano powder and other forms of graphene are made by a number of companies worldwide, mostly in the UK, the US, and China.

So far, we have only talked about the *mechanical* properties of graphene. It turns out that it has excellent electrical, optical, and thermal properties, i.e. it is a better electrical conductor than silver, which is otherwise the best electrical conductor known, and it has a thermal conductivity comparable to diamond, the best thermal conductor known. We will look again at the electrical properties in Chapter 6 when we look at novel transistors made out of graphene. Where do these electrical and thermal properties come from? In any conducting material, when a voltage is applied to it, the electrical charges (electrons) in the material move, and this is what we call an electric current. The electrons are held inside any piece of material due to electrical forces applied on them by the atoms. How strong those forces are determines how easily the electrons can move through the material. In fact, the atoms in materials are also held together by electrical forces and are the reason why things don't fall apart. Gravitational forces are very weak relative to electrical forces and only become noticeable for objects larger than around 100 km across. As a schoolchild, I saw that the ratio of the electrical to the gravitational force between any two electrons is 10^{42}. An interesting number as many of us know — it is the answer to the question of life, the universe and everything, as written by Douglas Adams in *The Hitchhiker's Guide to the Galaxy*. When I was an undergraduate at Trinity College, he came to give a talk to the philosophical society, and I plucked up the courage to ask him if he chose the number 42 because of this ratio of forces. He enigmatically answered "Perhaps it was". He told someone else he just liked the number as it is 6 multiplied by 7, and there are as many "42" stories as there are people who asked him as far as I can tell!

Going from a conductor to a semiconductor and to an insulator, the differences between these materials boils down to the fact that the electrons are increasingly strongly bound to the atoms going from a conductor toward an insulator. We describe the overall binding force on the electrons using a concept known as the *effective mass*, which is a measure of how light or heavy an electron *appears to be*. They appear to be heavier in materials with stronger bonds than those with weaker bonds. For the electrons in graphene, it was shown sometime ago that their effective mass is zero, which means that they can move through the graphene unhindered. Even though they appear to have no mass, this does not mean that they can travel at the speed of light — in fact they travel at around 1/100 of the speed of light, which is still around 100 million times faster than in a metal. On the basis of this, we would expect to stop the use of all metals overnight and just replace them with graphene, but there are a number of reasons why graphene just doesn't behave so well, and why it is simply not practical to get rid of metals in this way and change our manufacturing processes. A particularly useful application of graphene that has emerged throughout 2017 is its use in desalination. This is traditionally carried out using expensive thermal processes and can cost up to $1 USD per cubic meter of water. The cost is highly variable due to the volatility of the cost of energy. Using a simple membrane to filter water costs an order of magnitude less. The principle is beautifully simple — water molecules are small enough to pass through the carbon rings in graphene, but salt ions are too large, so they are very effectively filtered out using a simple membrane made of graphene. With suitable scale-up of the manufacture of these membranes, desalination will soon be possible across the world, both developed and developing, for a significantly lower cost than now (Figure 36).

Vast resources have been ploughed into researching and making use of the properties of graphene. The EU alone has spent billions of Euros as the US and other nations and large-scale research councils. A table showing the proposed uses of this material is shown below indicating that it has potential in battery technology, lasers, packaging, medical devices, and others. It is too soon to tell yet whether any of these applications will come to pass in any sustainable way. In most of the suggested cases, the graphene is being considered as a coating to improve the properties of some surface or another. In many ways, this is something that is already being done using diamond-like carbon (DLC). This is a layer of carbon, usually a few tens of nm thick, consisting of a random mixture of

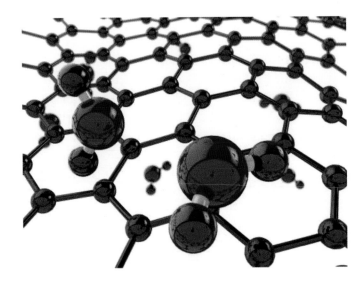

Figure 36.   Graphene is being used as a membrane to filter salt from water — for desalination. This is an emerging use of graphene that has gained a lot of industrial traction since early 2017. This is potentially a far cheaper process than traditional desalination. Here we show water molecules (red and blue) passing through a graphene membrane. Salt ions are larger and do not fit through.

crystalline and amorphous carbon, that is used in all manner of applications. It is used on turbine blades to reduce their air resistance and improve heat flow along the surface; it is used as a coating in engine components to improve their lifetime; on hard disk drives (HDD) to improve their durability and on medical implants for is low-friction and haemocompatibility properties. It is difficult to see how this can be replaced with graphene in a way that is sufficiently durable, but that is not to say it cannot be done. There are many examples of potential applications of graphene in fields ranging from battery and supercapacitor technology, sensors, oil extraction from oil wells, medical coatings, paints, inks and others, as shown in Table 1.

Nanowires: Nano in Two Ways

If we take a film and cut a thin strip from it, we end up with a *nanowire* or *nanofiber*. Fibers of materials have long been used for their advantageous mechanical properties. The most common example of this is carbon fiber — this consists of strands of carbon fiber, typically 5–10 microns in diameter, that are lighter and stronger than steel. These fibers are traditionally spun and then added to existing materials, fabrics, polymers, and even metals, to make them stronger and lighter.

The process used to make carbon fiber is expensive as it involves a lot of heating, but this manufacturing cost is dropping all the time. Carbon fibers were first

Table 1.   Applications of graphene.

Application area	Advantage	Industry affected
Batteries	Improves speed of charging and discharging, lengthens lifetime. Adding 0.1% by weight of graphene to the cathode can increase the above by up to 50%	Electric cars, mobile/ portable devices
Composites	Stiffness & strength of epoxy resins can be doubled by addition of as little as 1% graphene	Aerospace, F1 cars, wind energy, sports
Sensors	It has large surface area to volume ratio, high electrical and thermal conductivity and sensitivity to things on surface. Has been used to detect glucose, cholesterol, cancer, pH, gas	Medical, water and air testing, security, defense
Medical	Valve and stent coatings to reduce plaque buildup	Medical
Oilfields	Emulsion stabilizers, lubricants for drilling fluid; Enhanced Oil Recovery — add graphene nanomaterials to well to enhance oil mobility	Petrochemicals
Paints	Additive to paints as has anticorrosive effects	Coatings
Printing & Packaging	Conductive printing and packaging — is flexible, so can be bent, stable against humidity corrosion, temperature, printed electronics, such as RF–ID tags	Multiple
Supercapacitors	These are a battery-capacitor hybrid, useful for charging devices	Electronics
Transistors	Graphene is a superb electrical conductor and its versatility can surpass that of silicon. May be used to make transistors and integrated circuits. Scale-up is problematic	Semiconductor

developed in the 1870s by Edison who baked organic fibers (including cotton and spider silk) thereby carbonizing them, for use as lightbulb filaments — they weren't very good so the process fell into disuse (Figure 37). It wasn't until the 1960s that Rolls-Royce was one of the first companies to come up with a commercially viable means of producing carbon fibers and then using them, initially to enhance jet engine blades. In this case, the story wasn't all that rosy, as the blades were brittle, and impact with birds caused them to shatter. Later, the manufacturing process was developed to a point where carbon fiber could be made and then combined with other resin-based materials to form a composite material suitable for use in the fuselage of an aircraft. This was enabled as such composite materials are strong and light with excellent mechanical properties. Nowadays, composite materials are typically made using carbon fiber together with either a thermoset or thermoplastic polymer (Figure 38).

Carbon fiber-based composite materials are used in just about any application where their high mechanical strength and lightness are advantageous, and the price

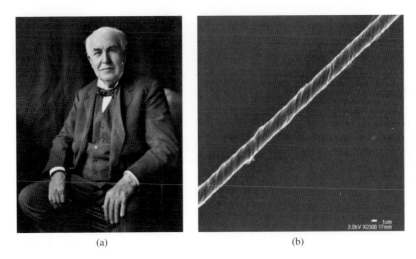

(a)   (b)

Figure 37.   (a) Thomas Edison, who first invented carbon fiber; (b) Carbon fiber — looks like a hair, but is around 10 times thinner.

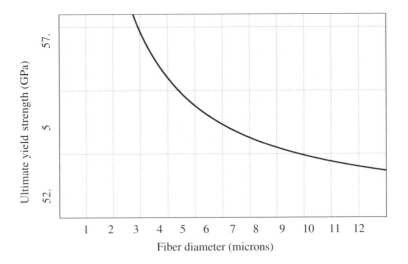

Figure 38.   Like many materials, the mechanical properties of carbon fiber are improved when they get smaller. This particular material plotted here levels off at around 1.6 GPa for very large fibers. For comparison, the ultimate yield strength for conventional structural steel is around 3–4 times lower at 0.4–0.5 GPa.

isn't too high. Examples include aircraft fuselages, car bodies (particularly Formula 1), crash helmets, military clothes, and a plethora of sports goods. In fact, over 1/3 of the carbon fiber in production today is used by the aerospace industry.

An example of how things have developed is the Airbus A350 XWB range, introduced in 2013, where 53% of the aircraft is made from composite materials (for comparison, the Boeing equivalent, the 787 Dreamliner comprises 50% composite materials). The composite material consists of carbon fiber reinforced polymer. The wingspan of 64.8 m and wing surface area of 443 m^2 is the largest single structure ever made from carbon fiber composite. This is a significant step up from the Airbus A380, which was introduced in 2005, and comprises a more modest 20% composite materials, with a significant amount of aluminium used in the fuselage. It is not possible to carry out a direct comparison between the two, but the weight of the empty aircraft divided by the maximum number of seats is 20% lower in the A350, with what appears to be around 100% better fuel efficiency. The use of plastics in this way enables the wing shape to be molded with greater ease, so more complex and ultimately more efficient shapes can be implemented. The wing tips are typically curved up in a fashion similar to a bird's wing to minimize wing-tip vortices in the air which create significant drag (Figure 39).

Wind turbines are another area where carbon fiber is used routinely (Figure 40). Although the rotor blades are usually hollow and made from fiberglass, typically blades longer than around 150 ft use a carbon fiber backing strip to add structural integrity and support.

By and large, the mass-produced automotive industry has not taken up with carbon fiber all that much. This is ultimately a matter of cost — it is still too expensive to modify existing tooling and fabrication plants to start molding carbon fiber, and the raw materials is too expensive given how low the profit margin for car manufacturers is. One exception is BMW, who make the entire frame for their i3 electric cars in a single piece, from carbon fiber (Figure 41). They start with an organic polymer that is heat-treated to carbonize it.

(a)   (b)

Figure 39.   (a) Airbus A350 XWB — the wings are the largest single structure ever made from carbon fiber composite, which can be molded to any desired shape; (b) A bird in flight. Note the upwards-curved wing tips which help to improve the aerodynamics.

Figure 40. A wind turbine uses carbon fiber to add structural stability to the blades.

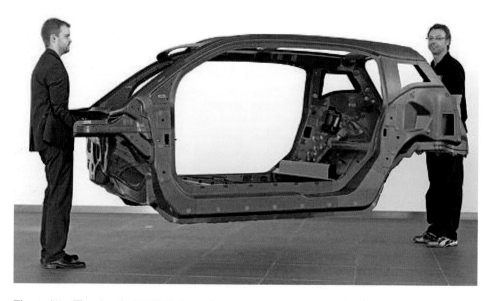

Figure 41. The electric BMW i3 frame is made from carbon fiber so it is very strong yet light. Given that the battery alone weighs around 200 kg — which is more than a typical gasoline engine, the overall weight of the car is comparable to a conventional one.

High-performance cars have been using carbon fiber for years now, particularly Formula 1, where frankly cost is not a barrier.

If carbon fibers are several microns across, then why bother talk about them in a book about nano-stuff? If we look a little closer at a typical fiber (in this case, with a diameter of 5 microns), we see that it comprises even smaller strands that are nanometer sized (Figure 42).

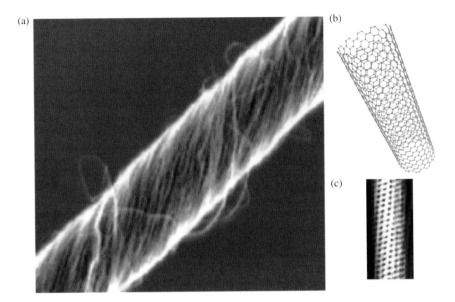

(a)

(b)

(c)

Figure 42. (a) Carbon fiber close up although the overall fiber is several microns in diameter, it consists of even smaller fibers only a few nanometers in diameter; (b) Atomic arrangement in a carbon nanotube; (c) STM image of a single carbon nanotube, diameter around 1 nm. The black dots are individual carbon atoms.

This is getting toward the ultimate nanowires: carbon nanotubes, or CNTs as they are more commonly called. As we saw in Chapter 1, CNTs have been around for quite sometime, but their official discovery is considered to have been when Sumio Iijima spotted them when looking at fullerenes he had made through a newly-developed process in an electron microscope in 1991. CNTs are made whenever any material is carbonized, or burnt at high temperature, for instance, whenever I bake anything. They are often found in soot, although the exact type of CNT and the quantity produced depends very sensitively on the heating conditions.

Strictly speaking, as CNTs are hollow, they are nano*tubes* rather than nano-*wires*. Consequently, they have a very low density — part of the reason for their high strength-to-weight ratio which makes them so desirable. Their discovery was hot on the heels of fullerenes, the most well known of those being C_{60} which we encountered in Chapter 1. The discovery within such a short period of time of two new forms (allotropes) of carbon, both at the nanometer-scale, led to a major surge in interest both in carbon itself, and in nanostructures in general, which in turn sowed the seeds that led to the development of advanced composites and nanomaterials. Many of the people who are working on graphene now started their scientific lives by working on either C_{60} and/or CNTs (Figure 43).

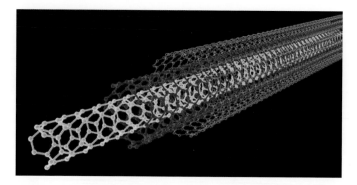

Figure 43.   Multi-walled carbon nanotubes are easy to produce, have high electrical and thermal conductivity and are very strong and have found applications in composites.

Carbon nanotubes, for a while, were considered to be the next great thing. They had quite remarkable mechanical and electrical properties, being stronger than steel per unit weight. A typical CNT has a diameter on the order of 1 nm, although it can be as small as half that, and can be up to several nm. Around the same time, it was found that concentric nanotubes also existed, where one nanotube was nested inside another. These became known as multi-walled CNTs as opposed to the simpler single-walled CNTs.

A significant amount of time, effort and funding was poured into research into CNTs, in the hope that they would provide us with better composite materials, better computer displays, and a range of gas sensors. The peak in this wave was in the early 2000s, just before graphene was discovered, and it has pretty much dissipated since then, as much of the initial promise simply could not be delivered, and there is evidence that CNTs are highly carcinogenic, even more so than asbestos. Notwithstanding this though, there are still applications where CNTs have been found to be useful particularly in the orient and the middle-east, where they are often used in composites. CNTs and carbon nanoparticles have however found somewhat of a niche in military applications, particularly in lightweight body armor (multiwalled tubes are used for this, so no worries about carcinogenic activity) and conductive paints for stealth aircraft. The stealth bomber in Figure 44, is coated with a paint containing CNTs and carbon nanoparticles that have just the right electrical properties. In order to absorb radar, the coating needs to not be too good a conductor (if it is, it will absorb and then re-radiate), nor an insulator (which won't absorb at all) — it needs to be *lossy*.

If CNTs were considered to be so great, then what went wrong? There was a combination of factors, along the lines of what we have already seen with nano in

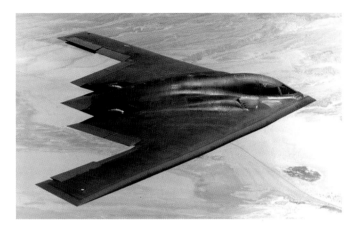

Figure 44.   The Stealth Bomber. The invisibility to radar is achieved through the use of a conductive paint containing carbon nanotubes and other carbon nanoparticles.

general, and in fact it is fair to say many areas of science suffer from this particular confluence of effects:

- Exaggerated claims from the beginning
- Material properties cannot be tuned at will
- Technology simply too costly to be scaled up for mass production.

In the case of the third point above, mass production of CNTs has increased at least tenfold over the past decade and is now at the scale of thousands of tons per year, so scale-up has been achieved. And they are just not used for their electrical properties. The main point for CNTs is that their properties just cannot be controlled all that easily as we cannot control their size accurately enough — a common issue with nanometer-scale things.

It is true that the mechanical properties of CNTs are excellent, and may be better than graphene. The properties that are important are: (i) the tensile strength and (ii) the Young's modulus. The tensile strength is the maximum stretching force per unit area (which we call stress) that can be applied to an object before it breaks, whereas the Young's modulus is the ratio of stress-to-strain (strain is the ratio of the deformation to the initial length). Tensile strength and Young's modulus are both measured in *Pascals* (Pa), which are the units of pressure — atmospheric pressure is 100,000 Pa. Tensile strength is usually around 1000–10,000 times smaller than Young's modulus, so we will quote values in GPa which are Gigapascals (1 billion Pascals) and TPa, Terapascals (1 trillion Pascals).

The bigger both of these values are, the stronger the material is. For comparison, we have tabulated these values for a number of common materials in Table 2,

Table 2.   Strength of materials versus cost and weight.

Material	Young's modulus (TPa)	Tensile strength (GPa)	Specific strength relative to stainless steel	Cost £/kg
Stainless steel (high carbon)	0.2	1	1	0.43
Graphene	1	130	420	60,000
Carbon nanotubes	1–5	10–50	312	30
Carbon Fiber reinforced plastic	0.29	1.24	3	1
Silk	2.2	1.4	45	1500
Copper	0.1	0.22	0.1	6.75
Kevlar	0.1	3.6	90	3,000
High-strength concrete	0.03	0.012	0.02	0.04
Wood (oak)	0.01	0.09	0.5	0.9
Polypropylene	0.002	0.03	0.15	0.2

so we can see just how strong graphene and carbon nanotubes really are. We have also tabulated the "specific strength", which is the ratio of tensile strength to density. This is a useful measure, as another factor generally important when manufacturing things is how much material is required and what its weight is. Rather than get bogged down in the messy units involved, we will use stainless steel as the reference and give it a specific strength of 1, so the other materials will be measured relative to it.

On the basis of these mechanical properties alone, it is not clear why concrete is so widely used, as it is not particularly strong, and is inferior to wood for many purposes. This is a compelling reason for using wood frames in houses rather than the environmentally unfriendly concrete and reinforced (with steel) concrete. Yet we know that concrete is the material of choice for many buildings and large structures. When we look at the cost per kilo of the above materials, the reasons become clear! Interestingly, the material which wins on the basis of the balance between strength and strength per unit cost is carbon nanotubes. Why then aren't we making our buildings out of this stuff? The answer is a rather mundane combination of the fact that we can't make CNTs in sufficient quantities and they may be highly carcinogenic.

Looking at the above data, the next best material is carbon fiber reinforced plastic, followed by stainless steel, polypropylene, concrete, and wood. The issue is that plastics are simply not as durable as concrete, steel & wood, their thermal characteristics are not favorable and their processing and tooling costs are high. As with most things in life, economics tends to dictate which materials are used for any given purpose.

We have mentioned silk, which is another example of a nanofiber, this time naturally occurring. The typical silk as spun by a garden spider, or orb-weaver (Araneus diadematus) is like a rope, consisting of many strands. Each of these strands is typically 100–200 nm in diameter, and there are generally 10–30 such strands in a single fiber, which is typically 2–5 microns in diameter. The strands are mainly formed from a complex blend of proteins, and a spider will produce strands with different morphologies depending on their intended function (Figure 45). Although spider silk has the most impressive mechanical properties of any naturally-occurring material, it is simply too expensive to be used in any real application. There are anecdotes that it has been used by Carpathian peasants to dress wounds, as it has antiseptic properties (probably due to its acidity), and they are high in vitamin K, which aids blood clotting. It has been used in optical instruments (microscopes, telescopes, etc.) to create fine cross-hairs for aiming. However, my very favorite story is that in 2009, after 82 people had been working for over 4 years to collect webs from over 1 million golden orb spiders in Madagascar (At this point, you, like I, are probably asking "*why?*"…), enough silk was collected to make the world's largest such piece of fabric, at enormous expense. It is now housed in the V&A museum in London (Figure 46). An arachnophobe's worst nightmare.

If spider silk is to be used more widely, then a better way of obtaining it apart from farming spiders needs to be developed.

This is where things take a turn toward the distinctly bizarre, in the same vein as growing replacement human ears on one's arm. Recent advances in the production of artificial spider silk have had mixed results. At the turn of the 21st century, a Canadian company genetically modified goats so that their milk would contain

(a) (b)

Figure 45.   (a) A garden spider & its web. Thankfully I have a good zoom lens on my camera — this was close enough for me! (b) Electron microscope image of spider silk fiber. These consist of multiple strands, which tend to be aligned.

Figure 46.   A cape made using spider silk. This is the largest single piece of cloth ever made using spider silk (*Image Courtesy*: The Guardian & the V&A Museum).

spider silk protein (spidroin). This was all well and good, but the ability to spin that protein into a fiber that has similar properties to spider yarns has proven to be a difficult challenge, and one not yet solved. The company that developed the technology (Nexia Biotechnologies) went bankrupt in 2009, but the work continues in several nearby universities, so this is something to keep an eye on. There has been more success in genetically modifying other organisms, ranging from silkworms to bacteria in order to make them produce silk. More recently in 2013, a German company succeeded in manufacturing spidroin from bacteria and then spinning it into fibers that are identical to spider silk. These are primarily being used in medical applications, for dressings, implant coatings, and cosmetics, aided by the fact that they are biocompatible and ultimately biodegradable. No doubt in the future if the process can be further scaled up, we will hear more about these fibers and see them used in a wider application space. This is yet another example of trying to mimic nature. It has had billions of years and evolutionary pressures to fine tune just about every aspect of life and its processes. It makes perfect sense for us to try to learn from it and mimic it wherever we can.

Nanoparticles: Nano Every Which Way

We have seen so far that 2D nanomaterials, i.e. thin films, are widely used and have a wide range of applications. This includes anti-reflection coatings on glasses, functional coatings on windows, electrical devices and via graphene, improvements of the mechanical properties of all sorts of things. The next step was nanowires and nanotubes, known as 1D nanomaterials, useful for their electrical and mechanical properties. The final step, you've guessed it, is 0D, i.e. tiny in all dimensions. Such entities are called *nanoparticles*, which we encountered earlier in this chapter, and we will come across again in Chapter 6. Nanoparticles are *everywhere*. Whenever we burn anything, it creates soot particles, some of which will be below 100 nm in size. Nanoparticles are widely used in a whole range of applications which we have touched on earlier, so we will look at this now in a little more detail.

The area where nanoparticles are most used is in cosmetics/personal care products. There are broadly two reasons for their continued use — (i) they are small enough that they can be absorbed rapidly by the skin and (ii) they offer better UV screening than larger particles. For the first of these, the form of the nanoparticles ranges from liposomes (like micelles that we will see in Chapter 7, but containing a tiny nanometer-sized droplet of water at the center), nanocapsules for controlled release, usually of proteins deep into the skin, nanosilver and nano-gold for their antibacterial properties. A number of products on the market use one or more of these properties together, for example L'Oréal revitalift, which is an anti-wrinkle cream uses polymer nanocapsules to deliver the active substances (Vitamin A — otherwise known as retinol) to deeper in the skin than larger particles can get. Alternatively, without needing to encapsulate the active ingredients, we can simply add the nanomaterial to a liquid that it does not mix with, in which case they will form a nanoemulsion (Figure 47). Common examples of emulsions containing nanometer-sized droplets are milk and some paints. This concept, leading to a large number of very small particles which by virtue of their size are more rapidly absorbed by the skin than larger particles, has been made use

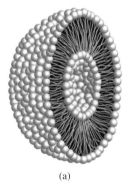

(a)   (b)

Figure 47.   (a) A liposome — An example of the sort of structure found in a nanoemulsion. This image is what a liposome would look like if cut in half. The white blobs represent the ends of the molecules, which are shown in green. Typical size in cosmetics is 20–200 nm; (b) One of the products that uses nanocapsules such as liposomes for enhanced absorption of active ingredients.

of by most cosmetics firms including L'Oréal (which has the most patents with nano in the title), Proctor & Gamble, Clinique, Neutrogena, Avon, and Estee Lauder. There are also a large number of small firms doing the same.

Should we be concerned about the use of nanoparticles in cosmetics? The simple answer is yes, as it is a complex issue with many compounding factors and there are many unknowns. The legal position is that cosmetics firms do not need approval to launch new products, but there is the understanding that they are responsible for their customer's safety and they do need to clearly indicate on the label if there are nanomaterials present. In other words, you would imagine that cosmetic firms will not knowingly release a product on the market that may not be safe. Of course, the problem is that they may not know whether something is safe or not as the tests that are carried out on cosmetics are just not rigorous enough. Manufacturers of course know what the ingredients are, and they can characterize the final product in terms of composition, particle size and distribution, and toxicity. They will have tested their products on a number of willing subjects, but it is often only after prolonged use that problems are revealed. Is the rise of cancer due to nanoparticles in cosmetics? The answer is we do not know, but the chances are probably not, that instead our risk of contracting cancer is due to a number of risk factors including diet and air and water quality, combined with longer lives and increased stress levels. Not a satisfactory answer, I know, but nonetheless, I do feel it is important that manufacturers run active research programs that look into these issues, or at least engage freely with academics who are interested. On balance, the sort of nanoparticles I am talking about are usually in the form of a liposome or micelle, so are purely a collection of molecules that happens to have nanometer dimensions. It is difficult to see how that could be dangerous in itself. The issue is more to do with *unintended consequences*, i.e. the small size of these structures and their enhanced uptake by the body means they may end up going where we don't intend. Unfortunately, there is no known way yet to see if this is actually happening or if it is a problem. Time will tell.

The other cosmetic-related area where nanoparticles are used is in sunscreen/ sun creams. Conventional sun creams use either zinc oxide or titanium dioxide particles, typically in the micron scale. These are also the primary active ingredients in other paste-type creams such as those we use for nappy rash or other rashes. Both types of material are *inorganic*, i.e. they do not contain carbon and do not come from or contain living matter. High protection factor sun creams use smaller particles than lower factor ones and have a significant number of what we can call nanoparticles, with diameters below 100 nm. The smaller particles are used as they have better UV absorption properties. The bulk of scientific literature that has looked into this has found that down to diameters of around 30 nm, the properties of nanoparticles are unlikely to be of concern from a health perspective.

However, below 30 nm, the novel properties of nanostructures due to their increased surface area and increased reactivity in general are a concern. We will touch on this again in Chapter 7.

Therefore, it would be beneficial for industry to take heed of academic findings and be more transparent, as it would set our minds at ease. An exciting development is that a collaboration between Chinese, Italian, and American scientists has successfully created artificial melanin with a high concentration of iron nanoparticles. The resulting material is in nanoparticle form, with diameters of the order 100 nm, and is therefore readily absorbed by the skin, is harmless and is extremely effective at blocking/absorbing UV rays (Figure 48). This kind of approach of recreating what biology already does, is one that science has tried to do in many areas with varying degrees of success. At the moment, we are in a position where there is rather a lot of misinformation regarding high factor sun creams, so we are in a catch-22 situation — use lower factor creams and definitely increase our risk of skin cancer or use higher factor creams and possibly increase our risk of cancer or some other condition brought on by the nanoparticles. I am not trying to scaremonger — quite the opposite, but in the absence of more robust information I would rather be cautious. The same goes for just about everything. Is butter good or bad for us? Red wine? Avocado? GM crops? The list goes on and on, and the need for sensible dialogue is becoming ever more important. It is unfortunate that by virtue of the way we receive news nowadays, it is easy to pick up alarming headlines and much harder to poke deeper to see where they come from and if they are baseless. Still, we must keep trying!

Figure 48.   High-factor sun creams tend to have a high proportion of nanoparticles due to their excellent UV-absorption properties.

Now that we have an idea of what nanomaterials are, where they are used and why we are so interested in them, it is useful to have a look at how we characterize them, which we will concentrate on in Chapter 5 where we look at the tools of the trade for imaging nanostructures and performing experiments at nanometer-scales. The insights that this have afforded us have been invaluable over the past 40 years and have informed us as to exactly what is going on with nanomaterials and why they behave the way they do.

Chapter 5

Seeing at the Nanoscale

I never met a man so ignorant that I couldn't learn something from him

— Galileo

Everything has beauty but not everyone sees it

— Confucius

Arguably, it was the invention of the scanning tunneling microscope (STM) in 1981 that kick-started the entire field of nanotechnology. Since then, a plethora of related techniques, each with its own acronym, have been developed. This family of microscopes is known as scanning probe microscopes (SPMs). Together with optical microscopy and electron microscopy, SPM affords us the ability to not only observe matter down to the atomic scale, but also to alter it (Figure 1).

Why We Need to See What Things Look Like

A detailed knowledge of what things look like allows us to better understand their properties and how they work, so imaging at the nanometer scale is an important

Figure 1.   The image that started it all for me: IBM has been "written" using single atoms and then imaged using an STM, 1989 (*Image Courtesy*: IBM).

area. It was an image taken with an SPM that got me interested in pursuing a career in this field in the first place. Until SPMs were developed, people had to resort to electron microscopy or guesswork to figure out what many nanostructures looked like. The reason why this is so important is that, as we have already seen, the properties of materials fundamentally change when their size is at the nanometer scale, and their shape also plays a crucial role. Therefore, in order to truly understand the relationship between shape, size and properties, we need to be able to obtain detailed images of nanostructures. The better the microscopes, the more we can learn.

To put this into context, we have seen in Chapter 2 when we looked into the developments of classical physics and how it evolved into modern physics, based on increasingly detailed and precise measurements revealing departures from expected behavior. We saw that early astronomers used their eyes and then crude telescopes to map the motion of the planets and stars. We then saw that as telescopes got better, they were able to see that the planetary orbits were elliptical rather than circular, and then that there were deviations from this which led to the discovery of new moons and eventually, the theory of relativity.

A step-change occurred with the launch of the Hubble space telescope into orbit in 1990, which has allowed us to explore our solar system and deep into space with unprecedented detail, so we now have a better understanding of the scale and structure of the universe (Figure 2).

At the end of the day, it's all about getting better images, and being able to see things more clearly. The same is true of small things — microscopes have improved to the point that we can now observe things at the scale of atoms. The tipping point with microscopy was when the STM was invented, as it opened up the gates to a vast number of measurement and characterization tools. I would like

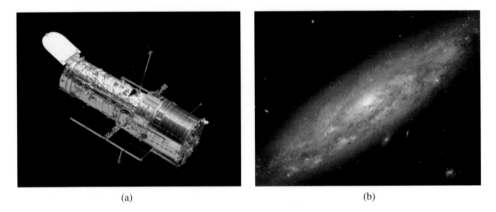

(a)   (b)

Figure 2.   (a) The Hubble Space Telescope, which has revolutionized the way we see the universe; (b) An example image taken with Hubble of the galaxy NGC 3972 (*Images Courtesy*: NASA).

to briefly consider the background to microscopy and how it has evolved over the centuries before delving into where we are now.

As seems to be the case over almost all important scientific discoveries, there is some disagreement over who actually invented the microscope. What we do know is that the name "microscope" was coined in 1625 by Giovanni Faber to describe a device made by Galileo in 1609. The word "microscope" comes from the two Greek words *micron* and *skopein*, which mean *small* and *to look at*, respectively. This is in contrast to the word "telescope", which comes from the Greek *tele* and *skopein*, meaning *far* and *to look at*, respectively. This word was also coined by an Italian (this time Giovanni Demisiani) to describe another device made by Galileo, also dating from around 1609–1610. There are contrasting claims that the microscope was invented by a Dutchmen, either Janssen or Lippershey nearly 20 years earlier, in 1590. Either way, we can say that the optical microscope was invented at around the turn of the 17th century (Figure 3).

The early microscopes suffered from a range of drawbacks, including the fact that the lenses, as they were generally quite small, were not of very high quality, so they were rather difficult to use. Like the telescope, a microscope needs to use at least 2 lenses in order to achieve a decent magnification. A one-lensed microscope, called a *simple microscope* can only offer low magnification of a few hundred times at the very best, and usually far below this. Examples of simple microscopes are magnifying glasses and eyepieces, which have been around for about 2,500 years — the first mention of the use of one (which was a glass vessel filled with water rather than a lens as we know it today) was in a play by the Greek playwright Aristophanes in around 424 BC. *Compound microscopes* comprise a magnifying lens placed close to the object and an eyepiece further away, the combination of which can achieve a magnification up to around 1000 times.

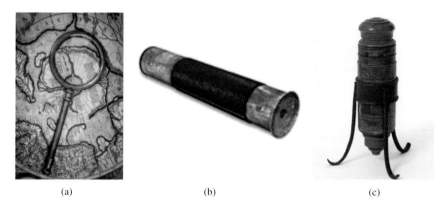

(a)   (b)   (c)

Figure 3.   The humble beginnings of the microscope. (a) A magnifying glass, otherwise known as a simple microscope; (b) Jansen's microscope, circa 1590; (c) Galileo's microscope circa 1609.

(a)   (b)

Figure 4.   (a) Antonie van Leeuwenhoek (1632–1723) — the first person to popularize microscopy and recognize the value in being able to see and study microscopic objects; (b) A replica of van Leeuwenhoek's simple microscope with magnification around 200 times.

The early microscopes, as popularized by prolific enthusiasts such as Antonie van Leeuwenhoek in Delft in the Netherlands in the mid 17th to early 18th century, were simple microscopes that managed to magnify items a few hundred times, which was enough to observe objects such as bacteria, cell structure, sperm cells, muscle fiber structure, and a variety of small plant and animal species. His work, particularly on protozoa, has led to him being considered by many to have started the field of microbiology. His microscope, as shown opposite, used a single, almost spherical glass lens, made using a process that he kept secret, ensuring that he had the monopoly on studying microorganisms at the time (Figure 4). He is said to have made around 500 lenses and 200 microscopes, some of which are still in existence in various museums around the world. The design was crude but simple and allowed him to move the sample relative to the lens to allow him to focus on different parts of it as required. He made detailed drawings of his observations and submitted them regularly, along with descriptions in the form of letters to the Royal Society in London. For his efforts, he was elected to a Fellowship of the Royal Society (FRS).

Over the next century or so, developments were made in the processing of glass that allowed better quality, higher magnification compound microscopes to be made by more people. One of the issues is that the magnification strength of a lens is directly proportional to its curvature, i.e. the smaller and more curved a lens is, the higher the magnification it produces. However, the smaller a lens gets, the harder it is to ensure that it is uniform and smooth, inevitably leading to distortions in the images it produces. Another issue relates to the fact that when white light passes through glass, it starts to spread out into its constituent colors, a process known as *dispersion*, as each color travels at a slightly different speed. When the corresponding rays recombine on passing from the glass back into air, they have become spread out, so do not all focus at the same point. This leads to smearing and discoloration of images. It was found that these issues can be partly overcome if two lenses are used, neither of which has a particularly high magnification, but together, they can do.

As mentioned earlier, when the compound microscope was developed, higher magnifications became possible, which opened up the ability to see more bacteria and the individual components in cells.

Microscopes and telescopes are ubiquitous — many are to be found in schools, and in teaching and research labs all over the world, and telescopes are the staple Christmas present for many a teenager, not to mention the odd keen parent using their kids as an excuse to get such an educational resource, myself included (Figure 5(a)). For some reason, my (it must be said, very sweet and patient) long-suffering wife refers to these essential learning tools as "toys", but when for a science project at school my 8-year old (at the time) daughter managed to take the image of Saturn (Figure 5(b)), the joking turned to acceptance. The toy had grown up and become a useful thing to have in the shed after all.

Things have moved apace with the development of digital cameras and CCDs, so now it is commonplace to connect microscopes and telescopes to imaging devices which are interfaced to a computer, and rather than have an operator

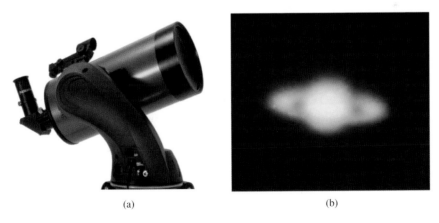

(a)   (b)

Figure 5.   (a) A typical high-end consumer telescope; (b) An image of Saturn taken by my daughter when she was eight. Let the curiosity live on!

look into an eyepiece, they can look at a computer screen and save the image for processing. One can now also do everything remotely, so you don't even need to be in the lab in front of a microscope or out in the cold garden with a telescope to carry out experiments by and large. There are an enormous number of different *types* of image that can be obtained using an optical microscope, depending on the exact information that one requires. For example, if one wants to measure the size of an object, then you can use an eyepiece with a graticule that has known markings on it, like a ruler. It is possible to image differences in materials by using filters, polarizers, and by measuring differences in reflectivity. One can immerse the lens in water, and perform measurements on cells, materials, and organisms *in vitro*, where they are still alive and functioning. It is possible to combine optical microscopes with spectrometers, to look at how materials absorb different wavelengths of light, and hence perform rudimentary materials analysis.

Nowadays, for microscopes and telescopes, the standard objective lens (the one nearest the thing you want to look at) is a compound lens, i.e. comprises several individual lenses. For example, in a typical high-end smartphone camera, the lens is formed from 5–6 individual lenses.

Going Beyond Light

However, despite this wide usage, there is a basic limit as to how well any optical microscope can resolve an object, as predicted by Ernest Abbe in 1873. This is known as the diffraction limit. This originates from the wave nature of light and is not dissimilar to Heisenberg's uncertainty principle that we introduced earlier. The resolution depends on the wavelength with shorter wavelengths (toward the blue end of the spectrum) showing better resolution. The bottom line is that even with the very best optical microscope, the maximum magnification that is attainable is around 1000. By immersing the sample and lens in oil, this can be pushed to around 1500, but it is generally not possible to go beyond that. In practical terms, this means that we cannot see things clearly once they are any smaller than around 200 nm. Now, for many cases, this is clearly not a problem, but for anyone wanting to explore the nanoworld, it is a showstopper! To try to resolve smaller features then, it is necessary to use shorter wavelengths of light, below the visible range which spans around 390–700 nm. However, trying to manipulate and focus non-visible light is a major challenge, and the constraints on lens design become even more stringent. There are ways around this, and a variety of techniques have been developed including passing light through tiny (10–100 nm diameter) holes which then act as an imaging aperture — this is a technique called near-field optics, and was the topic I spent 5 years working on as a PhD student and then for a postdoc position in Germany in the mid-late 1990s. This technique has demonstrated a resolution of around 10 nm so is very powerful, but it is very limited in what it can do, hence my reason for moving on to other, more fruitful things. Another technique, with the acronym STED (stimulated emission depleted) microscopy, was developed starting in the mid-1990s, and allows imaging of samples (usually biological) with a resolution of around 100 nm. The inventor, Stefan Hell, was awarded the Kavli Prize for nanoscience for this in 2014, and the Nobel Prize in Chemistry the same year, together with Eric Betzig, one of the early pioneers of near-field optics. In fact, it was Betzig's work in the topic that led to my wanting to do my PhD in the same area. While these techniques are impressive in what they can do, they are all rather complicated to use and cannot be used to resolve features/objects below around 10 nm. The solution was spotted by a number of researchers after de Broglie postulated that particles would have a wavelength, and Ruska invented the electron microscope in 1933. His design was so well thought out, that it is still used today. The basic concept was to focus a beam of electrons onto a sample in much the same way that an optical microscope focuses a beam of light onto a sample. The difference is that electrons cannot be focused by passing them through a glass lens, but they can be focused using a magnetic or electrostatic "lens". Passing a beam of electrons through a magnetic coil can be used to focus them.

As mentioned earlier, the wavelength of electrons can be fractions of a nanometer, so even though electron microscopes are also diffraction limited, their resolution limit can be orders of magnitude better than an optical microscope. The electron microscope has undergone a revolution in recent years with benchtop systems now hitting the market in droves, and they are becoming ever more affordable. Coupled with advances in automation, it is now possible for essentially untrained users to obtain useful images and data within hours, which is very useful within the industrial context. The principle of operation of electron microscopes is that a beam of electrons (taken from an electron source known as an *electron gun*) is focused down onto the surface of interest, and then depending on the type of microscope, this beam is either scanned across the surface to cover a square area and the transmitted or reflected electrons are collected, or the beam is held fixed, and the transmitted electron beam is expanded up again and projected onto a screen, thereby projecting an image of the sample. The latter is known as a transmission electron microscope, or TEM, and the former is known as a scanning electron microscope, SEM (Figure 6).

The highest resolution microscope is the TEM, which is capable of subatomic resolution, but it does suffer the drawback that the sample must be thin enough that electrons can pass through it, usually meaning less than 100 nm thick.

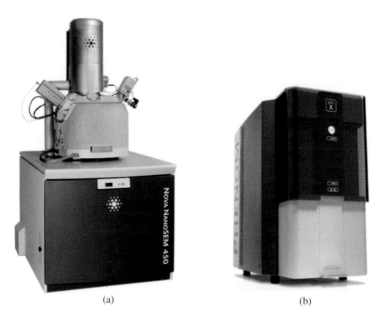

Figure 6. (a) Modern electron microscope — this is about 1 m × 1 m across; (b) Advances in min-iaturization have enabled the development of benchtop SEMs — this one is around the same size as a domestic coffee machine.

TEMs are rather expensive, typically at least $1M, so are not as commonplace as SEMs, which are typically $300–$500k. Put in context, an optical microscope costs anywhere between $10 and $100k, depending on how many bells and whistles are on it. The SEM typically has a resolution of around 1–3 nm. The advantages of the SEM are clear — simply by manipulating the electron beam's size and focus, we can image areas as large as around 1 cm, and as small as around 100 nm, with a continuous control over magnification. The depth of field of an SEM is far greater than that of an optical microscope and can be anything from a few mm to a few nm, so detailed images can be taken without needing to take too much care of the focusing. SEMs are usually quite easy to use, and someone can be trained to take basic images within a few hours. Of course, to get the very best out of any microscope can take years of training, but this is not necessary in many cases.

There are two big drawbacks of electron microscopes though, that mean they simply cannot be used to look at just everything. First, they operate in a vacuum, as otherwise the electron beam would just spread out too much passing through air, by hitting off air molecules. This vacuum is problematic when looking at organic matter, as it often causes structures to collapse on themselves or to swell, both of which result in damage. Second, when imaging, the sample is being bombarded by a beam of electrons. Now, those electrons need to go somewhere as otherwise, the sample will start becoming electrically charged. This would create an electric field around the sample that will deflect the incoming beam, resulting in unstable images. This effect is imaginatively known as *charging*, and to overcome it, samples need to be coated with a thin layer of electrically conductive material, usually gold or carbon. As you can imagine, this is fine for many samples, but not so for organic ones, such as cells, skin, hair, or thin fragile molecular films. There are ways around this, but the real point is that it is not possible to look at living specimens, or samples *in vitro*. Some examples of images taken with an electron microscope are shown in Figure 7. One of the drawbacks of both the optical and electron microscope is that they cannot easily give accurate depth information — the images they produce are 2D. There are exceptions to this in that there are special optical and electron microscopes that can give depth information and obtain 3D images, but these are costly instruments that are not trivial to use.

The Scanning-Probe Revolution

This brings us to the early 80s, when researchers tried to get the best of both worlds — very high resolution down to the nm scale and even beyond, with the ability to perform measurements under a wide range of pertinent conditions, ranging from ultra-high vacuum to in a liquid, and with vertical as well as lateral information, i.e. 3D images. This is where the scanning probe microscopes come

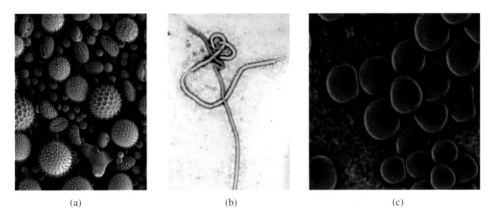

(a) (b) (c)

Figure 7.   What we can see using an electron microscope. (a) Pollen as seen through an SEM. The field of view is around 50 microns across; (b) Ebola viruses as seen through a TEM. The filaments are 60–80 nm wide (*Image Courtesy*: Dr. Frederic A. Murphy, CDC); (c) Staphylococcus aureus as seen through a scanning electron microscope. These are the bacteria that reside in your nose and are around 0.5–1 micron across.

 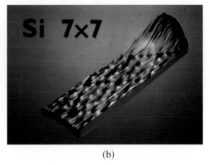

(a)   (b)

Figure 8.   (a) Heinrich Rohrer (left) and Gerd Binnig with the first STM; (b) The first image Binnig & Rohrer took of the silicon surface, showing individual atoms (*Images Courtesy*: IBM).

in. The STM was developed with the intention of understanding the structure of materials at the atomic scale, and after it was built and demonstrated that it could resolve atoms on known surfaces (gold was the first metal surface that was examined), the first problem it tackled was that of the surface of silicon, as there was a push from the semiconductor industry to do so. At that point in time, there were two conflicting models for the arrangement of atoms on the surface of silicon, and it was felt that the only way to resolve this issue would be to directly image the atoms. This work was carried out by IBM in Rüschlikon in Zurich, by Gerd Binnig, Heinrich Rohrer and their technician, Christoph Gerber (Figure 8(a)). In a matter of less than two years, they created the world's first STM and produced images showing individual atoms on surfaces (Figure 8(b)).

It is hard to overstate the shock wave that this sent through the scientific community — the first reaction was that other scientists either wholeheartedly accepted the results and saw that they must be true, or they assumed that the IBM group had made a mistake and were actually just measuring noise. Having worked with STM since 1991, I have seen many images that looked like neat little rows of atoms on a surface that turned out to be nothing other than noise due to lights, vibrations or other artifacts. However, with careful experimentation, these effects can be reduced almost completely, and true atomic resolution images can indeed be obtained. There is no doubt whatsoever that those initial images were real, which became apparent soon thereafter when others started to verify them. The biggest breakthrough at that time was when the STM was used to tell which way the atoms actually arrange themselves on that silicon surface, as it showed that one of the models was indeed correct. It was that experiment that led to the rapid acceptance of STM and explains why Binnig & Rohrer were awarded the Nobel Prize just a few years later in 1986, which incidentally, they shared with Ruska for his invention of the electron microscope. The STM is a precision instrument, capable of resolving differences in height of less than 1/100 of a nm (Figure 9).

Figure 9.   An early STM from IBM showing the tip close to a sample.

An image of the first STM is shown in Figure 8(a) with the inventors. Binning & Rohrer filed the patent for its design in 1979, and as mentioned earlier, produced the first atomic-resolution images less than 2 years later. The first image IBM produced of the silicon surface is shown in Figure 8(b). The label "7 × 7" refers to the particular arrangement the atoms take on the surface — they form patterns that are seven times larger than the spacing between the atoms underneath. Now, pretty much all metal and semiconductor surfaces consist of neat rows of atoms, which isn't all that interesting. However, what is important is to understand the role of *defects* in those surfaces, as they play an enormous role in determining the electrical properties of any material. The more defects a semiconductor has, the worse its electrical properties are. This is where the STM really comes into its own, as it can see atomic-scale defects. It is surprising just how many ways the atoms can misarrange on a surface — an example of an STM image that I obtained of a silicon surface with a variety of missing atoms, and single atomic steps is shown in Figure 10.

Until the STM came along, there was no way of visualizing this, or of knowing *where* any defects were.

The principle of operation of the STM is that a very sharp metal tip, which looks like a pin, but is much, much sharper, is brought to within around 1 nm or less of a conductive surface, e.g. a metal or semiconductor. A small voltage is applied between the tip and the surface, and a small current, of the order 1 billionth of an Amp, will start to flow between them. This current is a quantum tunneling current, which gets increasingly larger as the tip is brought closer to the surface, at an exponential rate. Therefore, small changes in the distance between tip and sample give rise to significant changes in the tunneling current. The tip or sample is fixed to a scanner unit which is capable of moving a few microns in all three directions, with a resolution of a fraction of a nm. The tip is then scanned over the

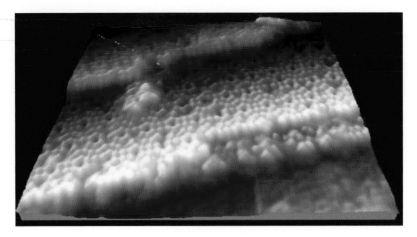

Figure 10.   An STM image of a silicon surface showing atomic steps and defects (where atoms are missing).

surface, the current is measured, and a feedback loop connected to the scanner ensures that the tip follows surface features in order to keep the current constant at some set value. The area to be imaged is split into a matrix of image points, and the distance by which the scanner had to move up or down to keep the current constant at each image point is recorded. This information is then used to generate an image of the surface topography. The resolution is determined by a number of factors, but the most important one is the sharpness of the tip. Most researchers make their own tips using either very sharp cutters or by electrochemical etching. The best STM tips are made either from tungsten or an alloy of platinum and iridium (Figure 11).

Shortly after the STM was revealed and images of silicon were shown to the world, others followed and started making their own variant of the STM. The first commercial system was produced in 1986 by Digital Instruments, Inc. and was called the "Nanoscope 1". This operated in an ambient environment, so was rather limited in what it could see. In order to be able to see pristine, atomically-clean surfaces, one really needs to house the microscope in an ultrahigh vacuum (UHV) chamber. To give an idea of the sort of pressure we are talking about, atmospheric pressure is defined as 1000 mbar. A typical UHV STM operates at a pressure of 10^{-10}–10^{-11} mbar. Under that sort of pressure, it takes around half a day for a surface to become covered with an atom-thick layer of contamination.

Bear in mind that all surfaces are covered with material that has condensed on them from the air and consists of a mixture of water and hydrocarbons. This can be several nm thick, so will get in the way if you are trying to see the underlying atoms on the surface. Therefore, to be able to see those atoms, the contamination

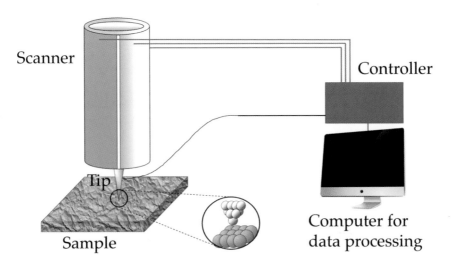

Figure 11. Principle of the STM. A very sharp metal needle, called a tip, is brought very close to a sample of interest. A small voltage (typically around 1 V) is applied between the two. When the distance between the tip and the sample is around 1 nm, a small current flows, called a tunneling current which is quantum in origin. This is constantly monitored as the tip scans over the surface, and the tip-sample distance is altered in order to keep the current at a constant value by means of a feedback loop. An image of the surface is then built up by showing how much the tip had to move up or down at each point. The tip is so sharp it has one atom at the end, and the tunneling current depends very strongly on the distance between the tip and the sample. Both of these lead to the microscope having very high resolution — capable of seeing single atoms. By moving the tip even closer to the sample, we can even move atoms around into any desired pattern. As the operation relies on a tunneling current, STM can only be used to look at electrically conductive samples.

needs to be removed. The most effective way of doing this is to heat the surface to a few hundred degrees to boil the contaminants off, but in UHV so that it then stays clean for long enough to take some decent images. An alternative is to carry out the experiments under a liquid environment, i.e. to immerse the sample and the last few mm of the tip in a liquid. The presence of the liquid can help to keep the sample surface clean. The first UHV STM was produced by Omicron in Germany in 1987 and set the stage for more detailed STM experiments across the globe. There are now dozens of STM manufacturers making STMs that cost anywhere from around $10k–$500k, depending on the level of complexity. The more expensive microscopes allow one to carry out measurements at very low temperatures, down to below 1K. There are many situations where low temperature is necessary, for example, to stop atoms or molecules diffusing around on a surface, or to measure weak electronic effects that would get smeared out at higher temperatures. Many of those pioneering scientists who were involved in the early days of STM passed on the expertize of how to build such microscopes themselves, and

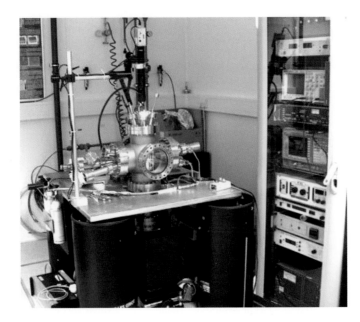

Figure 12.   The home-built ultrahigh vacuum STM from my lab. The actual microscope is the size of a soda can — the rest is needed for vibration isolation, to create a strong enough vacuum, and control and process electronics.

luckily, I was one of that second generation of microscope builders. An STM that I spent several years perfecting is shown in Figure 12. Some examples of STM images are shown in Figure 13 demonstrating the sheer power of this type of microscope.

Not Just Pretty Pictures

If the STM was only capable of taking atomic-scale images, it would still have revolutionized the way we do scientific experiments — the ability to be able to see things at that level of detail has added to our understanding of a large number of surface processes, including oxidation, melting, corrosion, and catalysis, as well as a host of other subtler effects. However, the STM can do far more than just take pretty pictures, as was beautifully demonstrated by Don Eigler and co-workers at IBM in Almaden in the US in 1989. They used the STM tip to not only image but *move* atoms around on a surface. The iconic image they ended up with was the IBM logo constructed from 35 Xenon atoms that had been precisely positioned, making the smallest ever company logo (Figure 1)! This image graced the front pages of many an international newspaper and certainly inspired countless budding young scientists, myself included. I have found a useful approach whenever

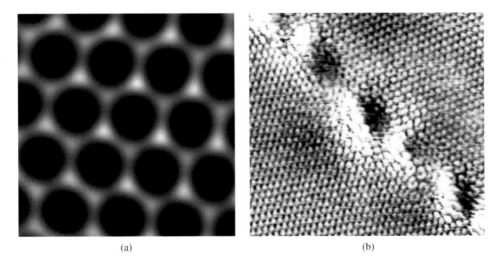

(a)   (b)

Figure 13.   (a) What graphene looks like in the STM. Each carbon atom appears as a bright dot; (b) A grain boundary in graphite, at the atomic scale. The bright spots dotted along the diagonal show that more electrons gather there than elsewhere.

Table 1.   Uses of atomic manipulation.

Science fiction	Science faction
Build molecules and other structures atom by atom	You can't build molecules atom by atom!
Carry out controlled chemical reactions one atom at a time	You can manipulate atoms into precise positions
The ultimate bottom-up fabrication tool	You can partially control the way atoms interact with each other
Create data storage devices one atom across	It's fun!
Spend vastsums of taxpayer's money to play with cool equipment and worry about the consequences after	Spend vast sums of taxpayer's money to play with cool equipment and worry about the consequences after. This one is just a joke. Isn't it?..Obviously

I get bowled over by some new technology is to remember what my grandmother said upon seeing the picture — *that's very nice, but what's it for?*. I found myself stumped, and rather than get cross because I couldn't think of an answer straight away, I came up with a list, which grew over time. It goes something as shown in Table 1.

The list is growing, but as with many things, the *direct* uses are not at all obvious apart from the fact that every now and again a paper appears on some STM experiment which does deepen our understanding of fundamental physical and chemical processes and is therefore immensely important and useful. Similar arguments are used to justify space exploration. Why should we go to the moon or mars, or further afield? Will we actually learn anything of any use to our daily lives in doing so? This is a question with a rapidly-evolving answer. A few years ago, I would have thought the answer was probably not. Now however, things are very different. With the thrust toward electrification of consumer and civilian vehicles, i.e. cars, we will eventually reach an impasse. This is down to a question of resources. The batteries used in electric cars contain significant amounts of cobalt, lithium, and graphite. Supplies of these materials are hard to come by and with the market growing, it is becoming ever more important to find alternative sources and materials. On the materials side, nanotechnologists are working on how to improve battery efficiency. Unfortunately, it is not possible to achieve the energy density in batteries that we achieve from hydrocarbons. The typical petrol contains around 45 MJ/kg, whereas the best battery has around 1/150 times that at less than 0.3 MJ/kg. So, to generate enough power for a car, we need big batteries, and there are ongoing efforts to source the raw materials from asteroids, the moon, and eventually mars. We should not overlook the fact that due to the space programs that various nations have run, the technology that had to be developed in order to get people safely into space and back again resulted in myriad developments in materials and communication technology, including Kevlar, LEDs, advances in freeze-drying of food (freeze-dried space ice cream is definitely an exception, not for the faint-hearted…), de-icers for airplane wings, and many other improvements. The advances brought to us by STM are less tangible, but they have helped to demonstrate that we have the ultimate imaging and positioning capability and I can testify to the fact that STM images are just cool and have inspired many a schoolkid to realize that maybe science isn't boring after all. As Feynman himself said, *Science is like sex — sometimes something useful comes out, but that's not why we do it…*

STM itself has proved to be an invaluable tool used to understand how molecular systems interact with the surfaces they sit on, and to better understand the electronic properties of surfaces. This sort of information is very useful to the advanced chemical industry and to academia in general.

I have already mentioned that an image demonstrating atomic manipulation was what got me interested in nanotechnology, and we have seen that iconic image spelling out IBM using 35 atoms. In those days, the word nanotechnology was not in common use, so I guess I was interested in the possibilities that moving single atoms around opened up. The first demonstration of moving single atoms in a controlled manner with an STM tip by Don Eigler was an endeavor fraught with challenges including (i) how to get atoms on a surface in just the right quantities that there weren't too many or too few, (ii) to keep the surface clean enough for long enough to move the atoms around and image the result, (iii) how to stop the atoms from moving around randomly, or diffusing on the surface, (iv) how exactly to do the manipulation, whether it is better to push or pull atoms with the tip. Through careful experimentation, these barriers were overcome. A key element is point (iii) above — to stop the atoms from moving about. There are two ways this can be done — either cool the sample and the tip to just a few degrees above absolute zero or use a surface that attracts atoms and holds them there strongly. Eigler plumped for the first option, as the second could have resulted in making it too difficult to move the atoms. Normally, cooling a sample and tip down in this way, combined with an ultrahigh vacuum would enable researchers to keep a sample surface clean for around a day. Eigler wanted to have more time to play with, so took the daring approach of cooling the entire microscope to just four degrees above zero. This resulted in surfaces remaining clean for *weeks* at a time. He settled on using Xenon atoms on a Nickel surface, as it was known that they would sit wherever they were nicely but would be possible to move without having to apply too much force. The atoms were moved one at a time, and after each attempt, the surface was imaged, just to make sure the atoms had been moved as planned. The experiment was carried out using a computer mouse to control the tip position, so was manual. Since those early days, atomic and molecular manipulation has become far more sophisticated. In Figures 14 and 15 is a series of images showing the sorts of structures that have been made in this way.

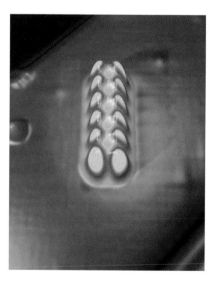

Figure 14.   An STM image showing 12 iron atoms that have been moved into precise locations to create a tiny bar magnet. Depending on the exact positions of the atoms, this structure can be made non-magnetic also (*Image Courtesy*: IBM).

Figure 15.   My all-time favorite atomic manipulation image — this is just 17 nm wide! (*Image Courtesy*: IBM).

In an attempt to get schoolchildren interested in nanotechnology, in 2013 IBM created a video, posted on YouTube, entitled "A boy and his atom". They carried out a sequence of manipulations of 65 carbon monoxide molecules on a copper surface over a two-week period, and the resulting 242 images were put together to make a stop-motion film. The image area was only 45×25 nm, and this video is now within the top 1% of most-watched entries on YouTube. Some of the stills are shown in Figures 16 and 17. Myself and many others have shown this video to schoolchildren, including my own, from the ages of 10–18, and all find it to be highly entertaining and more importantly, inspiring. I would go so far as to say that it has inspired a renewed interest in science among some children, as it shows that it can also be fun. It is particularly timely as well, as the STM and AFM are now mentioned on the A-level Chemistry and Physics syllabus, as well as on equivalent ones across the world, and therefore anything which can help to bring them to life is helpful.

A Microscope for Everyone — The AFM

While the STM was being taken forward, Binnig filed a patent in November 1985 for a related instrument — the Atomic Force Microscope (AFM). Together with Gerber, and Cal Quate at Stanford, who had previously invented the scanning acoustic microscope, he developed the first AFM. This microscope is far more powerful than the STM as samples don't need to be conductive, it can operate in

Figure 16.   The poster for IBM's "A boy and his atom movie", 2013 (*Image Courtesy*: IBM).

air, liquid and vacuum, and it can be used to map many surface properties apart from just simple topography. As such, the AFM is far more prevalent than the STM, with several thousand microscopes in operation worldwide. Many university research labs in materials science, surface science, chemistry, physics, engineering, biotechnology and others routinely use AFM to image and measure samples. In recent years, AFMs have found their way into industry and are used routinely by the semiconductor industry for quality control of silicon chips, and for general surface finish analysis in the coatings industry. Many sectors in

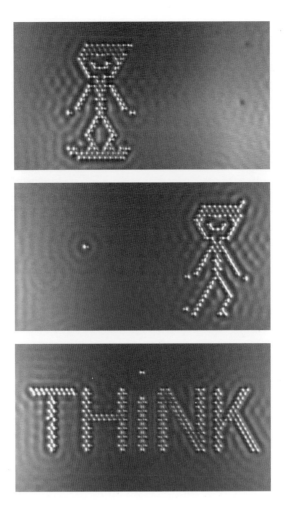

Figure 17.   Some of the stills from IBM's "A boy and his atom movie", 2013. The total area used for this was 45 × 25 nm (*Image Courtesy*: IBM).

manufacturing use AFM either directly or via university partners, to explore how to improve their products (Figure 18(a)).

The basic principle of operation of AFM is similar to that of the STM, with the exceptions that the probe is different, and the interaction between it and the sample is different. Apart from that, the scanner and the software are similar. In the AFM, the probe tip is mounted on a micro-fabricated cantilever beam, such as in Figure 18(b). It is a bit like a stylus record player in that the probe scans over the surface and follows the surface features. One of the joys of using AFM is that generally, it just works — one can insert a probe tip and a sample and start getting meaningful images straight away. With STM, it can take some fiddling about to get good images, but when they are good, they are excellent. In almost

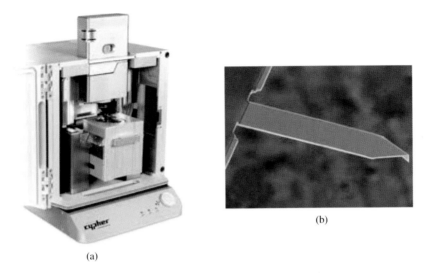

(a)

(b)

Figure 18. (a) A commercial AFM (*Image Courtesy*: Asylum Research); (b) AFM probe with imaging tip at the end (on the right-hand side).

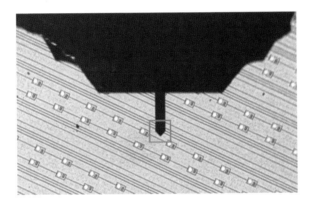

Figure 19. A view into an AFM's optics, looking straight down onto the probe, showing the probe, the sample, and the sort of area typically imaged (red square). Field of view here is ~1 mm.

all AFMs available today, a simple optical microscope is used to view the sample and the probe tip, and there is the capability to move the tip to any desired location on the surface, aided by the microscope. A typical view is shown in Figure 19, where an AFM probe is pictured on top of a device structure. The extent of the area that the AFM can image is indicated. The combination of optical imaging and AFM is indeed very powerful, as we will see later, particularly in biology.

The tip is situated at the end of a micro-fabricated cantilever beam, which is fixed at the other end. A laser is shone on the cantilever near the tip end, and the

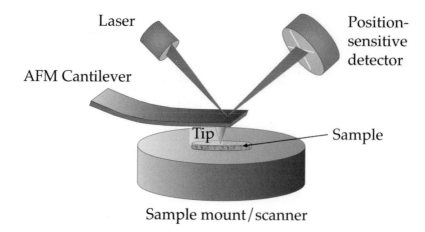

Figure 20.   Principle of the AFM. The basic principle of scanning a very sharp tip over a surface and monitoring an interaction between the two to build up a picture of the surface using a feedback loop is similar to the STM. The specific operation of the AFM is that the interaction is mechanical rather than electrical, i.e. we measure forces rather than a current. The tip is mounted on a cantilever beam, and as it scans over the surface, we monitor its position using a laser and photodetector. A laser is shone onto the end of the cantilever and a photodetector is used to collect the reflected light. As the cantilever bends when the tip moves to follow surface features, this causes the reflected light to move, which we measure with the photodetector. The major advantage of the AFM relative to the STM is that it can be used to look at samples that are not electrically conductive.

reflected light goes to a position sensitive photodetector. As the cantilever bends when scanning over a sample surface, the reflected beam position changes. This is measured using the photodetector, and the resulting signal is used for feedback control, to generate an image of the surface (Figure 20).

   A question that I often get asked is "so what can you look at with an AFM"? A good question of course, and the answer is *anything, as long as it's not too big or too small.* By too big I mean anything rougher than a few microns or wider than around 100 microns, and by too small, I mean anything smaller than an atom. For everything in between, it's perfect! We should consider some examples of the application of AFM to get a better feel for what it can do and why we should be interested. Most AFMs operate under ambient conditions (i.e. they do not require a big expensive and complicated vacuum chamber), and display a resolution of 1–10 nm, so although it is not good enough to see individual atoms, in most cases that isn't necessary anyway. Rather than randomly show images, I would rather think about different sectors and look at how the sort of images one obtains from AFM can be considered useful. If we start with applied work then, the biggest and arguably most important field of research in terms of potential impact on our daily lives is medicine. AFM is used by medical researchers, although it has not managed to penetrate the field as much as expected, at least not yet. Imaging of live cells was reported as

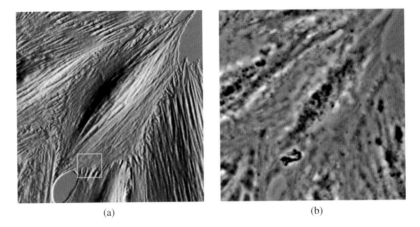

(a)   (b)

Figure 21.   An example of the integration of optical and scanning-probe microscopy. They see different things and are highly complementary. (a) Fibroblast cells as observed with an AFM. Image size is 90 microns; (b) The same area as seen with an optical microscope.

early as 1992, but bear in mind that AFM can only see what is happening on the surface of cells rather than within the interior, so to be truly useful it must be considered as a complementary technique rather than one that can do everything. Nonetheless, the surface can tell us quite a lot about what the cell is doing inside. For example, the Figure 21(a) shows an AFM image of fibroblast cells.

Fibroblast cells are of particular interest as they are the most common cell type in connective tissue and play an important role in wound healing. The overall image size is 90 microns, which is quite an enormous scale for most AFM users, and the extracellular collagen matrix produced by this type of cell is seen as the straight lines (filaments) emanating from it. We know that when there are several such cells close together, they create a structural scaffold matrix that eventually goes on to become connective tissue. When the same area is observed using optical microscopy, the filaments are not visible even under the highest possible magnification (Figure 21(b)). This highlights the typical scenario: optical microscopy is useful for getting an overview and can see internal structures (for transparent objects), but suffers from low resolution, and some features do not show up at all. In order to reveal internal structures, quite often it is necessary to stain samples. Scanning electron microscopy is also useful for an overview and can show the 3D nature of objects. TEM is capable of the highest resolution and can also see internal structures, but like SEM, it requires complex sample preparation techniques and sectioning (Figure 22). Finally, the AFM cannot reveal internal structure, but can see the gross morphology with very high resolution and does not require any complicated sample preparation. As with optical microscopy, the cell can be viewed

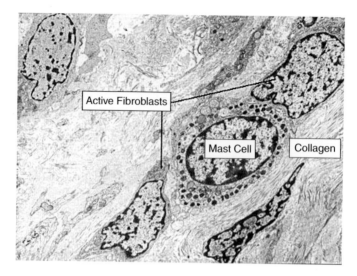

Figure 22.   Fibroblast cells as seen with a TEM. This gives the highest detail but requires complex sample preparation and an expensive microscope.

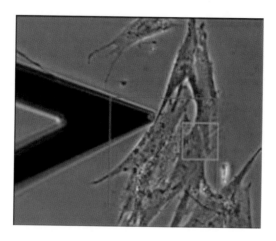

Figure 23.   Combined optical microscopy and AFM. The AFM tip is indicated by the red dot, and the AFM can be programmed to image any desired area, as indicated by the red and green squares.

in vitro, so does not need to be dead to be imaged. Many researchers will use a combination of optical and atomic force microscopy to study cells, as that way you can simultaneously employ conventional optical tools such as fluorescence alongside the high resolution of the AFM. AFM control software is now sophisticated enough that it knows where the AFM tip is and one can point and click on the area of interest to obtain an image (Figure 23).

Some other examples of biological samples that have been investigated by AFM are shown in Figure 24.

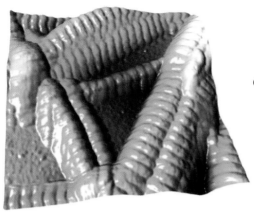

(a) Collagen. 5 micron area (*Image Courtesy*: Asylum Research).

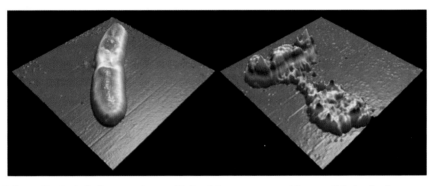

(b) Salmonella. Left: In its natural state. Right: After treatment with an antibiotic. 5 micron area (*Image Courtesy*: Asylum Research).

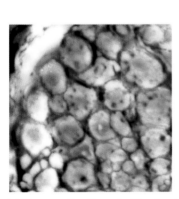

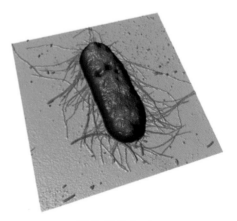

(c) Cross-section through skin showing individual cells. 25 micron area.

(d) *E. Coli.* 5 micron area.

Figure 24. Biological samples investigated by AFM.

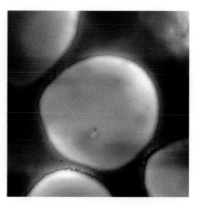

(e) Butterfly wing showing periodic structures. 20 micron area.

(f) Red blood cells. 12 micron area.

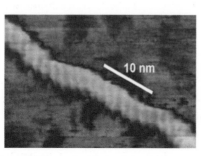

(g) DNA molecule — the pitch of the double helix shows up as stripes.

(h) Muscle fiber, primarily made up of actin filaments, 20 micron area (*Image Courtesy*: Derek Milner, University of Illinois).

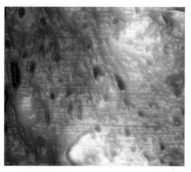

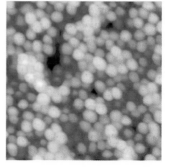

(i) Enamel on tooth surface after exposure to a fizzy drink for 10 minutes — lots of nano-sized cavities have started to form. Image size 1 micron.

(j) Staphylococcus Aureus — The bacteria in your nose. 10 micron area.

Figure 24.   (*Continued*)

What about timing? I have not mentioned that these images take time to obtain. For example, using standard equipment, it is possible to take an image with an optical microscope in real time, at video rates (over 25 frames per second). The electron microscope, both TEM and SEM, can typically capture a high-resolution image in a few seconds, although the sample preparation time and the time taken to wait for the microscope to be ready can be half an hour or so. By comparison, most AFMs will need around 10–20 minutes to take an image of this size. This was considered by many to be a showstopper, as quite often biological processes happen on the timescale of seconds or minutes, so they will be missed if an AFM is being used. There are situations for which this is not an issue, for example, when imaging red blood cells, which are essentially static.

For this reason, the top-end AFMs of today are now capable of obtaining images at video rates. This makes it possible to observe all manner of processes ranging from growth of crystals to cavity formation in teeth, to interactions between antibodies and antigens and many more examples.

So, AFM can tell us about the size and appearance of nanostructures in much the same way that other microscopes can, but with the added advantage of being able to measure their height and surface structures. To see how this is useful, Figure 25 shows the same structure (an electronic device) as imaged in an optical microscope, electron microscope and finally an AFM. The level of detail is clearly highest in the AFM. The AFM has a much higher spatial resolution than is apparent from this image, and it only really becomes noticeable when we zoom in to the micron scale and below. We can use the AFM to observe the grain structure within materials, and to distinguish between materials at the nanometer scale.

For example, a piece of gold, such as in my wedding ring, if looked at in an optical microscope, there is not a lot of detail to see apart from the odd scratch and dent. However, in the AFM, a whole wealth of detail becomes visible. Individual grains that were pressed together in the forming process can be seen and

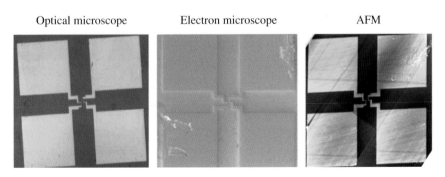

Figure 25.   An electronic device as observed using optical, electron, and atomic-force microscopes. Each image is 80 microns across.

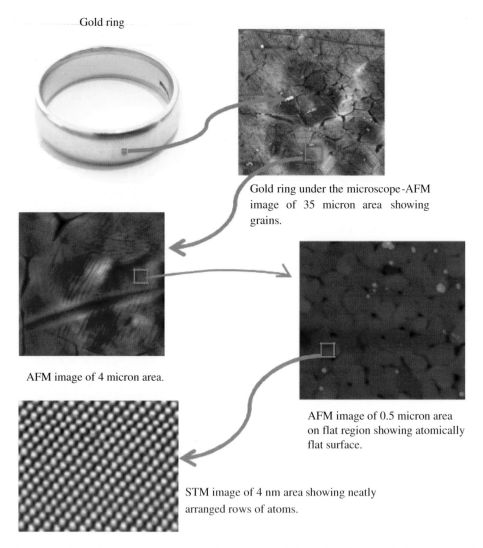

Gold ring

Gold ring under the microscope-AFM image of 35 micron area showing grains.

AFM image of 4 micron area.

AFM image of 0.5 micron area on flat region showing atomically flat surface.

STM image of 4 nm area showing neatly arranged rows of atoms.

Figure 26.   Put a ring on it — an example of structures at different length-scales, in this case a gold ring, all the way down to the atomic scale.

distinguished. Looking even closer, finer details of the surface can be seen — at the scale of a single grain, some will even have single atomic steps. This level of detail is beyond what most electron microscopes can see and is routinely observable with the AFM (Figure 26).

Another example is shown in Figures 28 and 29 — household aluminium foil. Foil is formed by casting aluminium ingots or billets and passing between two rollers. Generally, sheets are rolled two at a time, and the surfaces in contact with the rollers become shiny, whereas the foil surfaces that were in contact with each

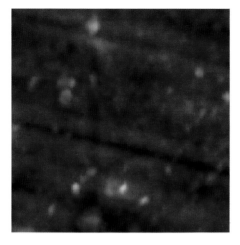

Figure 27.   AFM image revealing individual lubricant particles (bright spots). 1 micron area.

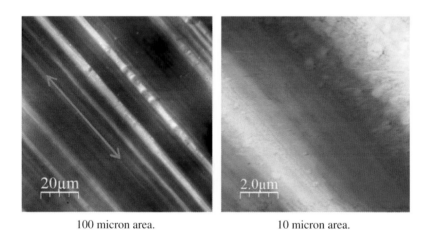

100 micron area.   10 micron area.

Figure 28.   Shiny side of aluminium foil as seen in the AFM.

other become matt, or dull. A small amount of lubricant is applied to the rollers to aid the process and prevent tearing. When we look at foil in the AFM, the effect of the rolling process and trace amounts of lubricant can be seen, as shown in Figure 27.

The roller markings can easily be seen both at large and small scales as shown in Figure 28. The image on the left is of a 100 micron (0.1 mm) area, and clearly shows the major grooves that are left by the roller and are around 1 micron thick and 10 microns apart, and show the rolling direction, as indicated by the arrows. The image on the right is of a 10 micron area in between some of the larger grooves.

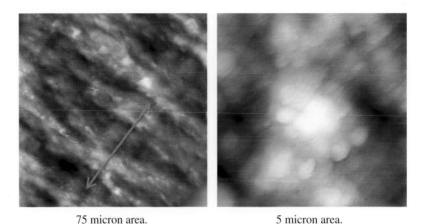

75 micron area.   5 micron area.

Figure 29.   Dull side of aluminum foil as seen in the AFM.

On the dull surface, the rolling direction is much less obvious, as this is the side that was on the inside, in contact with the other piece of foil, so was not in direct contact with the roller. The overall roughness is similar, at around 1 micron, but is essentially random, which is what leads to the matt appearance. The particulates that are observed on this side were mostly transferred from the rolled side. For packaging and distribution, the foil is rolled up on a cardboard roll, and the matt surfaces are in contact with the shiny surfaces, spreading the remaining lubricant particles onto both sides of the foil.

The Ultimate Power of the AFM

Now, this is all very well, and it certainly looks as if plenty of people have been joyfully fiddling about in their labs (myself included it must be said) looking at anything and everything with their AFMs to take these pretty pictures. If that was all AFM could do, then of course it would still be an invaluable tool. However, the real power (drumroll please…) of AFM is that it can do far, far much more. It is not particularly suitable for manipulation of objects as the STM can do with atoms and simple molecules, but it can take images of the *properties of materials*. What on earth does that mean I hear you ask? Well, as an example, topography alone tells us a great deal, we can determine the shape and size of things, but it cannot tell us *what they are*. The topic of chemical mapping is something we will touch on later, but for now we will concentrate on what AFM can do to map the properties of materials. In a nutshell, we can use it to see *variations* in at least the following properties of materials as shown in Table 2.

The list is growing all the time. We will spend the next few pages taking a dizzying trip through some of the many useful things we can use an AFM to do.

Table 2.   Properties that can be probed and *imaged* using AFM.

Hardness/softness
Stickiness
Friction
Chemical properties
Electrical resistance
Electric charge
Magnetic properties
Temperature
Thermal conductivity

Why is any of this important? Well, an example is food packaging, which is usually made from some polymer or another. When polymers are mixed together, to somehow improve their properties, we create what is known as a *polymer blend*. Quite often, the polymers become *phase separated*, i.e. they do not actually mix, and you end up with regions of polymer A surrounded by regions of polymer B or *vice versa*, a bit like what you get when you add oil to water and shake it — you get oil droplets surrounded by water. It is important to know to what extent the polymers mix, so that the proportions or components can be tweaked accordingly.

As it happens, when polymers mix like this, the regions they form tend to be in the size range of around 100 nm–10 microns, perfect for AFM to probe. Other forms of microscopy usually struggle to see the different areas, as the different polymers are often the same color (Figure 30).

Another example that is shown is our old friend, a fibroblast cell. As mentioned earlier, with AFM we can only image the surface, but with materials property mapping, we can see what is happening *beneath* the surface. We have shown an optical image of the AFM probe near a cell in Figure 31(a), and there is the AFM height image (topography), with the stiffness superimposed on it as a color map in Figure 31(b). Looking at this, we can see that some areas are stiffer than others, and by comparison with the optical micrograph, it is clear that these areas are (i) the nucleus and (ii) actin filaments. So, in a roundabout way, we can determine what's inside.

Another example is stickiness or *adhesion*. In the example shown in Figure 32, a blend of two common polymers, polystyrene (used in packaging) and polybutadiene (used in car tyres) is investigated using adhesion force mapping. In this mode of imaging, we bring the AFM probe into contact with the surface at each image point, and then measure the force needed to pull it back off the surface, which is called the *adhesion force*.

Figure 30.   Polymer blend. The different colors show three different polymers making up the blend, and are areas with different stiffness, as measured using AFM. 15 micron area (*Image Courtesy*: Asylum Research).

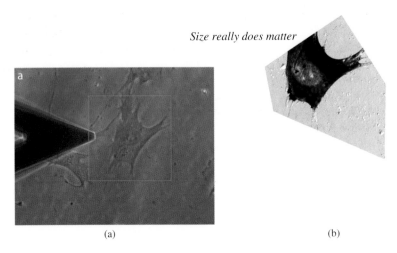

(a)   (b)

Figure 31.   (a) Optical microscope image of an AFM cantilever near a fibroblast cell. The green square is the area that was subsequently imaged by AFM; (b) The same cell as imaged by AFM. The color map shows the stiffness, with bright corresponding to stiff areas. The central, oval shaped area and the bright lines of increased stiffness are the regions around the nucleus and on actin filaments (*Images Courtesy*: Asylum Research).

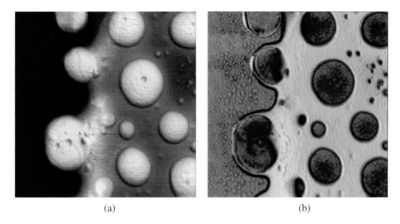

(a)   (b)

Figure 32.   (a) Topography image of a polymer blend, of polystyrene and polybutadiene (rubber used to make vehicle tyres). 4 micron area; (b) Adhesion map of the same area — the different polymers are clearly distinguished. The dark areas are polybutadiene.

Why did I choose this example? Well, this particular blend is pretty useful. On its own, polystyrene is actually quite brittle. By contrast, polybutadiene is rubbery. You wouldn't want to make any sort of container out of a rubbery substance, and also probably not out of a brittle one either. By adding a small amount of polybutadiene to polystyrene, we end up with a material that can bend more without breaking, as the bending stress is absorbed by the rubbery particles, many of which happen to be smaller than 1 micron. In fact, the image above shows that the polybutadiene particles are *polydisperse*, i.e. have a wide range of sizes, and a significant number are around 100 nm across. The resulting mix is known as *high-impact polystyrene* and is used to make toys and external casings. Again, using the AFM to understand how the polymers mix and the uniformity of the mixture is important at the manufacturing stage as it can be used to see if the production process is working effectively or if it needs to be tweaked.

Chemical structure of polystyrene

Polybutadiene:

Polystyrene container   Polybutadiene in car tire

Polystyrene + polybutadiene = high-impact polystyrene (HIPS)

Uses of HIPS:

- **Plastic Cutlery** — Needs to be strong and a little flexible.

- **Medical Trays** — This particular one shown here is for storing medical instruments, and has silver nanoparticles embedded in it which give it antimicrobial properties.

• **CD Cases** — The back is usually made from HIPS, as it needs to flex whenever a CD is inserted or taken out. The front is usually made from standard polystyrene, as it is glassy and clear so we can see the CD inside, but it is brittle.

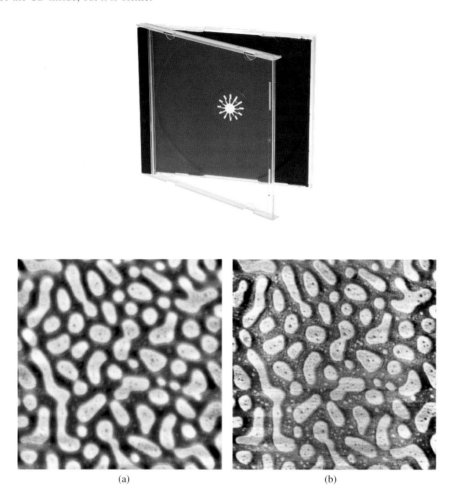

(a)   (b)

Figure 33.   Distinguishing between materials at the nanoscale. (a) Topography image of a polymer blend, 5 micron area; (b) Friction image of the same area. The difference in friction is how we can tell which material is which as one of the polymers has high friction (bright areas) whereas the other has lower friction (dark areas).

It is also possible to distinguish between materials due to their having different friction properties, as shown in Figure 33 of another polymer blend, this time comprising three polymers. From the topography image on the left, it is not possible to unequivocally distinguish between the different polymers, but there is a noticeable difference in the friction, as shown on the right.

As well as using the AFM to spatially map mechanical properties, we can also use it to map electrical ones as well. This is now used by the semiconductor industry in fault-finding on chips. As with any manufacturing process, there needs to be some form of quality control. Whenever a new process is being developed, each stage needs to be carefully monitored for flaws. Traditionally, the chips would be examined in an electron microscope whilst they were working to see if there are any breaks or short-circuits, but this is a slow process. The trend in recent years has been to use automated AFMs for checking silicon wafers at critical stages during the fabrication process. An example of such an instrument is shown in Figure 34. The need for automated AFMs is driven by the fact that in general, AFMs are not trivial to use and require significant user training to both obtain and then interpret measurements meaningfully and if a process is automated, it can run overnight, and one can come in in the morning to an automatically-generated report. I'm just glad my PhD students haven't asked for one of these beasts...

The primary use of this is in fault-finding in working circuits. AFM has sufficiently high resolution that it can see where any cracks are or where the voltage or resistance are not what they should be.

This sort of information is really quite useful, as a detailed knowledge of *where* wires break allows us to better understand the more important question of *how* they break. Once we understand that, we can design around it. Ultimately, it is generally due to manufacturing defects resulting in a wire that is too thin somewhere, which renders it fragile. When an electric current flows through any material, it causes damage — the electrons knock into the atoms in the wire as they flow, and eventually, this leads to breakage. This process is known as

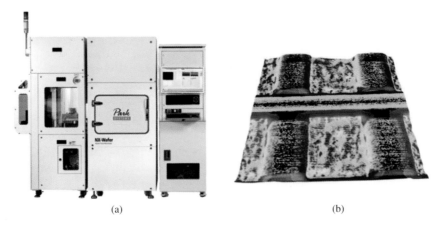

(a)   (b)

Figure 34.   (a) A fully-automated AFM for fault-finding in semiconductor circuits; (b) An image of a memory cell from a RAM chip. The brighter areas indicate regions of lower resistance. Scan area 5 microns.

electromigration and is a major headache for all device designers. By using AFM to locate where wires break, we have been able to build more sophisticated models for electromigration that actually agree with experiments, and this sort of information is then fed back into device design to create more robust circuits.

On the topic of electrical measurements with AFM, it is also possible to map electric charges on surfaces. As well as that, as we have exquisite control over where we place the AFM tip, we can also *inject* or *induce* charge at any desired location on a suitable insulating surface to build up patterns, along the lines of what we saw with the STM, albeit at a somewhat larger scale. Some examples of this are shown in Figure 35, where we have used an AFM tip to induce charge in a ferroelectric material, of the sort used in flash memory devices.

If we use a magnetic AFM probe, then we can use it to map magnetic fields from samples, and this can help in our understanding of magnetism at the nanometer scale. This technique is imaginatively called *magnetic force microscopy*, or MFM.

The magnetic properties of materials change once they are smaller than around 1 micron, and MFM can be used to gain insight into this. An example of the use of MFM to look at data on a computer hard disk is shown in Figure 36. The information is stored in a thin film of magnetic material in the form of *bits*, which are rectangular areas where the magnetic field of the film has been made to point in the opposite direction to the surrounding area. This is done using the *read*

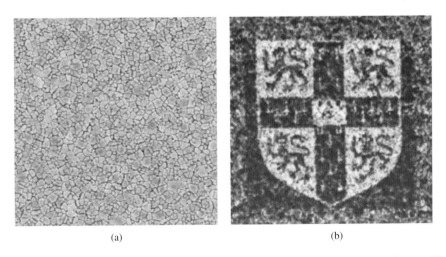

(a)   (b)

Figure 35.   (a) An AFM image showing a ferroelectric material that is polycrystalline; (b) Simultaneously acquired image of the surface charge showing that charge has been added in a controlled manner using the tip to, in this case, "write" the University of Cambridge's coat of arms. Image size 12 microns.

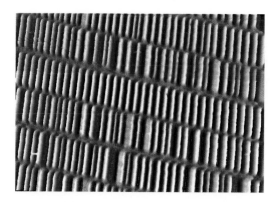

Figure 36. A MFM image of data on a conventional computer hard drive. The image area is 80 microns wide.

head. MFM has enough resolution to see these bits and can even be used in extreme circumstances (i.e. if you have enough time to spare) to actually decode the data on a corrupt disk.

Using AFM to See Atoms and Beyond

So far, I have just concentrated on the types of information and experiment that we can do with a bog-standard AFM operating under ambient conditions. The real beauty of AFM however, is that, like STM, the resolution is limited by the tip size, and it should therefore be possible to observe single atoms. This was some time coming, as there are so many different forces acting on an AFM tip at any one time, to be able to pinpoint those coming from single atoms is not trivial. However, in the late 1990s, developments in the electronics that control AFMs enabled atomic resolution, as shown in the next 3 images. As with the STM, in order to be able to obtain these sorts of images, the surface being imaged needs to be atomically clean, and therefore needs to be in ultra-high vacuum.

Figure 37 demonstrates that in fact, the AFM has a resolution very similar to that of the STM, at a fraction of a nanometer.

As well as being able to look at atoms and see how they arrange themselves in materials, wouldn't it be really amazing to be able to see the bonds that join those atoms? This was one of the holy grails of SPM technology that was finally realized in the summer of 2009. Researchers discovered that if they attached a carbon monoxide (CO) molecule to an AFM tip, they were suddenly able to see the bonds joining atoms together in molecules (Figures 38 and 39). The bonds appear as bright lines about 1/20th of a nm wide. It is possible to observe different types of atomic bond as well, including hydrogen bonds between molecules.

Figure 37. AFM image showing silicon atoms.

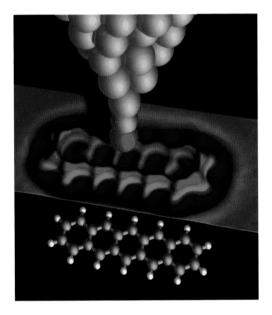

Figure 38. AFM tip with a CO molecule attached at the apex. The C atom is black and the O atom is red. The tip is just above a molecule which is sitting on a single-crystal surface. This sort of tip "functionalization" has made it possible to image atomic bonds.

These are the type of bonds that hold water molecules together and are responsible for the shape of proteins and for the double-helix structure of DNA.

So, what have we learnt from all of this then, apart from seeing pretty pictures? Well, for a start, we have seen that it is possible to *see* what materials really look like, all the way down to the atomic scale. We can move atoms and molecules around pretty much at will. We can probe the properties of materials, mostly their

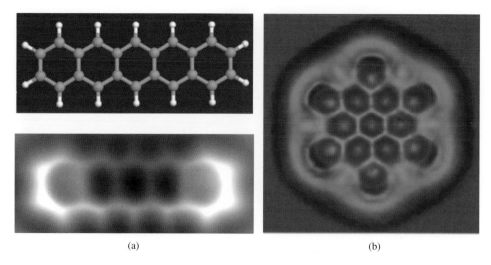

(a)   (b)

Figure 39.   (a) A single Pentacene molecule. Top: Traditional ball and stick model. Bottom: AFM image taken using a CO-modified tip. The bonds are seen as the bright lines; (b) An AFM image of a single molecule, revealing that the bonds (the straight lines) are not all the same length (*Images Courtesy*: IBM research).

surfaces, but in some cases, what is happening inside as well. Microscopy has truly evolved and reached the ultimate limit. Now that sounds very grand, but it is true!

Not only can we perform interesting experiments in the lab now, we can explore the properties of materials that are actually relevant to our everyday lives.

In practice, these microscopes are commonplace. Advanced SPMs and electron microscopes can be found in any University Chemistry, Physics, Materials Science, Engineering, and in some cases, Biotech departments, alongside optical microscopes which are so cheap, many people have some version of one at home. Why do so many people want to use them? It is part of the human condition — seeing is understanding, and the more detail you can see things with, the more you can understand them and how they work. There was a period in the late 1990s when it seemed that SPMs would remain niche tools that would never make it out of university labs and into the "real world", but advances in electronics and automation have enabled microscope manufacturers to create systems that do not need too much user intervention, and that can perform multiple measurements simultaneously. This greatly facilitates their use and increases throughput, making them a viable and desirable option for R&D and quality control purposes as well as for basic research. These microscopes have now evolved to the point where it is now even possible to automate atom manipulation using tricks learned from AI and self-driving automobiles. They have even found their way onto the syllabus for schoolchildren learning about chemistry and physics, and that it is perfectly reasonable to expect to be able to actually see what materials look like at the atomic level. In my day, that was unheard of. Things have come a long way in a short time — to my mind that is the measure of success of a technology that it permeates into our lives in positive ways such as this. The developments within the past few years since 2010 or so in particular in the imaging and measurement of biological systems at the cellular and molecular level have been extraordinary, and these are starting to impact medicine in real ways. High-speed AFM now allows us to monitor the chemical activity in cellular machinery at the nanoscale and with millisecond resolution.

Chapter 6

Nanotechnology in Electronics

There is no reason anyone would want a computer in their home

— Ken Olson, founder of Digital Equipment Corporation, 1977

Computers in the future may weigh no more than 1.5 tons

— Popular Mechanics, 1949

With the benefit of hindsight, these statements seem preposterous, especially given that the number of personal computers in the world by 2017 was 2.4 billion (*source*: Forbes), or around one computer for every 3 people. That's just computers, and if you look at the GSM Association mobile connectivity index which tracks the number of mobile phones in the world, it reached 8 billion in mid-2017, outstripping the number of people on the planet, which currently stands at around 7.4 billion (*source*: Population reference bureau). At the time the second statement above was made, computers were few and far between, mainly used for military purposes, and weighed in at several tens of tons, as we will see later when we take a look at the ENIAC and Colossus computers.

Nowadays, many of us take our technology for granted, and have multiple electronic devices which enable us to remain *connected*. I, for one use a desktop computer for most of my work activities, and then a laptop when giving talks or lectures, a tablet for general media browsing at home, a phone for all the above when I'm on the move, and of course a wearable device. It means that I never get any nasty surprises, and always know what is going on, or at least I no longer have an excuse for not knowing. In our house, with two adults and two children, a quick scout around revealed that we have three computers, four mobile phones, three tablets, various internet-enabled home devices, and counting. All in moderation, this ultimately makes work and life easier and more efficient, but it also means that

Figure 1.   Some of today's machines that we are slaves to.

there is a general blurring around the edges between work and not-work, which I suspect is not too healthy. Every time a new device comes on the market, we certainly have a look, as we are slowly and inexorably becoming more and more dependent on obtaining information, music and media, etc., in this way. I could not function without my smartphone's calendar and I can now ask the latest voice-activated addition to our kitchen what meetings or appointments I have on any given day. Its all good fun, and actually is enriching. At least that's what I keep telling myself anyway… (Figure 1).

The Incredible Shrinking Transistor

What has made all of these things possible is advances in the design and manu-facture of arguably the most important electronic component of all time — the transistor. This is what is at the heart of these devices and is what all electrical appliances use in some form or other, so developments in the electronics industry have far reaching consequences for almost all aspects of daily life, even in ways of which many of us are not aware. It is worth our while looking at this for a bit as the story of transistors is in many ways at the very heart of nanotechnology, and they are now true nanostructures. The development of the transistor started in the late 19th century with the likes of Edison and Tesla among others experimenting with vacuum tubes — partially evacuated glass tubes containing a small amount of a gas or vapour, and an electrode at each end and electrical properties that we can do something useful with. Edison patented his findings but diverted his energy into other things such as inventing a working lightbulb and devising and setting up

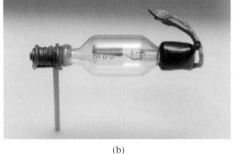

(a)   (b)

Figure 2.   (a) The first vacuum tubes invented by Fleming in 1904 — the precursor to the modern-day diode (Image from John Ambrose Fleming, 1919); (b) One of deForest's Audion devices from 1906 (Image taken by Gregory F. Maxwell).

the US national grid, so he did not take his ideas on vacuum tubes very far. This was however picked up a few years later in 1904 by Fleming who saw that the electrical characteristics displayed by some vacuum tubes, of allowing electrical current to flow in one direction and not the other, would be useful (Figure 2(a)). He refined their operation and is now credited with the invention of the vacuum diode, which was used by Marconi as the detector in his early radios.

Marconi had been experimenting with less efficient counterparts from the mid-1890s, building on earlier work by Hertz, the first person to demonstrate that electromagnetic waves could be generated and transmitted over some distance in the late 1880s, a result which he said was very interesting, but was probably not of much practical value. This was a mere 15 years after Maxwell predicted the existence of electromagnetic waves on the basis of his mathematical theory of electricity and magnetism. Such a pace of development is unprecedented in science, to go from an interesting and abstract theory to full-scale commercialisation and implementation in around 20 years. One of the issues with early radios was that they were difficult to hear as the signal strengths were rather weak. The search was on for some way to increase or amplify this signal without having to increase the transmitted power by much, as the transmission antennas were already being powered with so much electric current that they needed to be cooled. This led to the invention of a new kind of device — the vacuum transistor (otherwise known as a triode) for radios by deForest in 1906 — he called it the "Audion" (Figure 2(b)). As the figure shows, this looks more like a lightbulb than what we would consider a transistor. These transistors enabled signal amplification for the first time and their mass-production made it possible for enthusiasts to build their own radios. This rapidly led to a plethora of designs for radios which continued for the next few decades (Figure 3). The first licensed radio station went on air in the US on November the 20th, 1920, followed by many others and eventually

Figure 3.   An early radio, showing the vacuum tubes (Image from Wojciech Pysz).

Europe caught up when on the 14th of November 1922, the BBC made its first broadcasts, followed a few years later by the Japanese who started broadcasting on March the 22nd 1925. Of course, as is still the case today, there were plenty of *unlicensed* broadcasters in pockets all over the world before any of the licensed ones came about.

While all of this was going on, others were working hard at seeing what else could be done using electronics and transistors. Using the idea that electronic signals contained information, the next step was to carry out some sort of manipulation or *processing* of that information, in other words, *computing*. The high-level difference, of course, between a radio and a computer is that a computer has memory, so can store information rather than just decode it live which is what a radio does. It was not long until an electronic computer was developed, where information was stored in another component called an accumulator, which was a vacuum-tube-based register, not exactly the same as the sort of memory computers have now. Must is a great master as most of this happened precipitously during the Second World War. It started with the Atanasoff–Berry computer which was conceived in 1937 by Atanasoff from Iowa State University and tested in 1942 and was used to solve linear equations but suffered from reliability issues. The British codebreakers at Bletchley Park, using a design devised by Thomas Flowers with some input from Alan Turing, secretly developed a much more powerful machine, codenamed Colossus (Figure 4(a)), which was used to help decipher the Enigma code used by the Germans to send tactical military information between army units. The first machine went into operation at the beginning of 1944.

A series of Colossus machines were commissioned by the military and are credited with helping the allies to win the war. In fact, it has been said that they

(a)   (b)

Figure 4.   (a) The Colossus computer — UK built (Image from the National Archives); (b) ENIAC — the "Electronic Numerical Integrator and Computer" — US built.

would not have stood a chance without these machines. This project was kept a secret until the 1970s when GCHQ released some photographs of a Colossus machine and was only officially confirmed by the government in 2000. Unfortunately, this meant that many of those who worked on this project and played a pivotal role in the war never received the acknowledgment they were due. At around the same time as Colossus was being developed, the US developed the ENIAC computer (Figure 4(b)), first built at the University of Pennsylvania and fully operational from early 1946 and was used to calculate the trajectory of ballistic missiles. Its computing power was such that it could carry out calculations in 30 seconds that took a physicist 20 hours to work through. There is no question that both it and the colossus changed the course of history, and as computers they were far more powerful than humans, but they were expensive, big and heavy, used an enormous amount of power and were terribly unreliable. For instance, ENIAC cost around $6,000,000 in today's terms, it weighed in at around 30 tons, used 130 kW of power and only worked around half the time. By comparison, an iPhone 6 weighs 129 g, uses less than 1 W of power and can carry out roughly 154 million times more calculations per second. The architecture of the iPhone, and modern computers is of course quite different to the early computers, but the biggest step forward came with the development of the semiconductor transistor and the integrated circuit. Strictly speaking, the ENIAC and Colossus computers did not have transistors or the same kind of memory as we have today, so a direct comparison is not possible, but the vacuum tubes they contained have a similar function to today's transistors.

By comparison, ENIAC had just under 17,500 vacuum tubes, and the iPhone XS has around 7 billion transistors in its main processing chip, the A12 bionic. What has enabled this exponential hike in the number of transistors is that they have continually been getting smaller, as we will explore shortly. The trigger for all of

(a)   (b)

Figure 5.   (a) l-r: Bardeen, Shockley & Brattain from Bell labs in 1948 after the public announce-
ment of their invention of the transistor; (b) A replica of the first transistor.

this was the invention in late 1947 of the semiconductor transistor in Bell labs in
the US by Bardeen, Shockley & Brattain, which was far superior to any of the
vacuum devices mentioned thus far (Figure 5). It had the edge in terms of speed
of operation, low power consumption and low voltage operation, and earned its
inventors the Nobel Prize in 1956. For the next almost 20 years, radios, televi-
sions, and other electronic goods were made using these transistors which were
individually made. The first transistor was around 1 cm in size and was made
from germanium as opposed to silicon which is used today — an inspired choice
as (a) it contains around a thousand times more charges that can carry electric
current and (b) that current can flow about three times faster. The reason why we
use silicon then? Cost — it is around six times cheaper than germanium and is
easier to work with. It did not take long after the invention of the transistor for
the next development step which was combining several components on the
same piece of semiconductor, — the integrated circuit.

The idea of having discrete, i.e. separate components which needed to be
somehow connected together was a major barrier to creating computers with
increased computational power. To make more powerful computers, we need
more transistors, which requires more connections. Before the integrated circuit,
they all had to be made by hand, by soldering. This problem became known as
the "tyranny of numbers". Even ENIAC had half a million soldered connections,
which were notoriously unreliable. Faced with this problem, it became clear that
if there was some way that metal wires could be incorporated on the same piece
of semiconductor as the transistors and other components, then that would
greatly simplify fabrication and facilitate the creation of circuits with increased
complexity and reliability. The resulting integrated circuit was invented by a

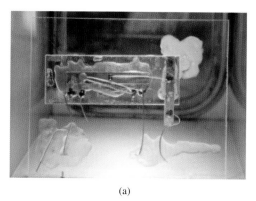

(a)   (b)

Figure 6.   (a) Replica of Kilby's first integrated circuit (Image by Florian Schäffer (Own work) [CC BY-SA 4.0 (http://creativecommons.org/licenses/by-sa/4.0)], via Wikimedia Commons); (b) Inside a computer where there is a processor (under the fan), memory chips, and slots for a variety of peripherals.

number of individuals independently at around the same time in 1958 and is usually ascribed to Jack Kilby of Texas Instruments and Robert Noyce of Fairchild semiconductor (Figure 6(a)).

The integrated circuit took off after that, especially when Noyce joined forces with Gordon Moore in 1968 to form Integrated Electronics, or INTEL for short. They pioneered the commercialization of the microprocessor based on their integrated circuits and the rest is history (Figure 6(b)). They have also pushed the limits of fabrication to be able to make semiconductor and metal structures with precision of the order 10 nm, and have helped to drive several branches of nanotechnology, as from the mid-1990s, the feature sizes on their chips started entering the nano-realm where many of the effects we encountered in Chapter 2 start to make themselves known.

One of the key developments that enabled integrated circuits to be made was the invention of optical lithography — akin to photography. This is a relatively straightforward process whereby a light-sensitive polymer material is deposited on a surface, in this case, silicon. A pattern is then projected onto that surface using a UV lamp and a mask, and wherever the light hits the surface, it degrades the polymer, which can then be removed easily using a suitable solvent. The unaffected polymer remains on the regions of the surface that were not exposed to light and protect the silicon. We now have a surface with polymer patterns on it which are a copy of the pattern on the mask, and these features can be as small as 20–30 nm. The silicon is exposed in the regions where the polymer was removed, and a whole series of processes can now be run such as metallization, doping, etching, etc., on that silicon. When complete, the remaining polymer can be removed using a different solvent to before. By doing this many times with different patterns, transistor, resistor, and capacitor structures can easily be built up layer-by-layer in highly complex 3D circuits. Advances in chip manufacturing have gone hand in hand with advances in lithography. The main developments over the past 40 years have been geared toward being able to continually increase the number of transistors that can be made at one time. This has been achieved in two ways — first, the lamps used to project light onto the silicon surfaces have been improved. One of the issues is that as feature sizes are now so small, and smaller than the wavelength of light, we get interference and diffraction of the light as it bends around the patterns on the mask. This is overcome by placing the mask in contact with the surface being patterned and by using light of shorter wavelength. The lamps used now emit light in the deep ultraviolet range, known as extreme ultraviolet, or EUV. The second development is that better lenses have been made which can uniformly project the light from the lamp over larger areas of the surface, which means that larger silicon wafers can be patterned than ever before. Initially, the wafers used in the late 60s and early 70s were 2 inches in diameter, whereas since mid-2017, they are now just under 18 inches. Having larger wafers simply means that a single machine can produce more circuits, increasing output. What is important is that these transistors are used to make electricity do something useful, in devices.

Phones and computers are forms of device. The word "device" gets bandied about a lot these days, and it means different things to different people. In its most general form, *device* means *tool*. By that definition, we can consider just about anything as a device, e.g. a pen, a hammer, a computer, a bowl, a phone, etc., and I am sure you can see where this is going. In techie circles, a device usually means a phone or an electronic gadget. A nanodevice specifically means an electronic *thing* which is (i) smaller than 100 nanometers across and (ii) whose properties are somehow different to larger-scale versions of itself — i.e. it must be a *nanodevice* rather than a nanometer-*sized* device. As it happens, this immediately rules out rather a lot of things I could be telling you about the microelectronics industry which has the largest number of patents involving the word "nano". This does not mean that the components in our computers are made using nanotechnology, it just means that many of them have some dimensions below 100 nanometers. We touched on this earlier when we talked about the rise of the electronics industry, but it is worth taking a bit of time to raise our awareness of how it all works, so we will spend some time now looking into the details of what is happening inside a computer.

All electronic components operate on one basic principle: using electricity to do something useful. That can mean moving something (motors), generating heat or light, sensing (temperature, strain, pressure, the presence of particular materials, etc.), detecting signals (radios, digital cameras, etc.), storing information, and processing information. Inside a computer or a smartphone, the same basic things are happening — information, or "data" is stored digitally as 1 s and 0 s in memory chips, and is then accessed as required and used accordingly. Each 1 or 0 is imaginatively called a "bit". This is all rather abstract, and the way in which data is actually stored is mostly in the form of electric charge from electrons (except for hard drives, where it is stored in a magnetic field). This charge is physically in a circuit component called a capacitor, to which a transistor is connected. The way in which this combination of components works is that the transistor will only conduct electricity if there is charge on the capacitor, and if this is the case, this is called a "1". No charge on the capacitor leads to a transistor which does not conduct electricity, and this is a "0". The data is written and read using wires connected to the transistor (the source and bit lines) and the capacitor (the word line). This concept is used in many contexts in different forms of memory ranging from USB memory sticks, to RAM, solid state memory and others, and is becoming more and more prevalent. The typical lateral size of each memory "bit" is around 300×300 nm, leading to a data storage density of around 1 Terabit per square inch, comparable to that in a magnetic hard disk (recent developments should result in ~10 times higher density available to the market within the next 2–3 years — according to Seagate) and far greater than for a DVD which stores around 45 times less at 2.2 Gigabits per square inch (Figure 7).

As a trillion-dollar a year industry worldwide (*Source*: Quora), this is big business, and there's no doubt that the continued developments in computers over the last 50 years have been driven by the market and its potential. This is best illustrated by the well-known "Moore's law" (Figure 8), a trend that has held roughly true since around 1965, when Gordon Moore from INTEL noted that the number

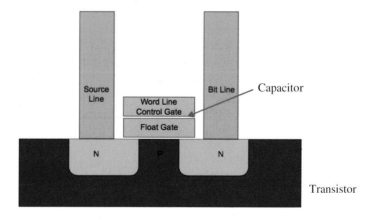

Figure 7.   Structure of a flash memory device (*Source*: Cyferz at Wikipedia).

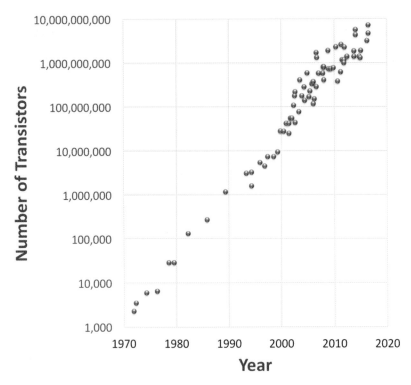

Figure 8.   Moore's Law — The number of transistors per microchip has been steadily increasing over the last almost 50 years. (*Data Courtesy*: G. Lisensky, Nanotechnology FYI).

of components on integrated circuits was doubling every year, and he predicted that this would continue for at least the next decade. He revisited this in 1975 and predicted on the basis of technological advances at the time that this doubling would occur roughly every 2 years. This again held true until around 2006 when processor development started speeding up, and it is only within the past 1–2 years that the slowdown has started.

Each time a new processor is designed, a new process to make it has to be developed (Figure 9), which has to be capable of making smaller features each time this happens. This involves building a new factory each time, at a cost ranging from $1–10 billion. Currently in development (as of the beginning of 2017) is the 10 nm process — this does not mean that features are 10 nm across, they are more like 40 nm, but the number refers to the minimum half-pitch between neighboring devices on the same chip. Chips made using this process have been released in consumer devices by Samsung in 2017.

The next generation process is the "7 nm" process, which was commercially released in 2018, and this will be followed by the "5 nm" process processors in 2020. This trend of shrinking transistor size is illustrated in the Figure 10, which

Figure 9. Development of transistor processes since 2005.

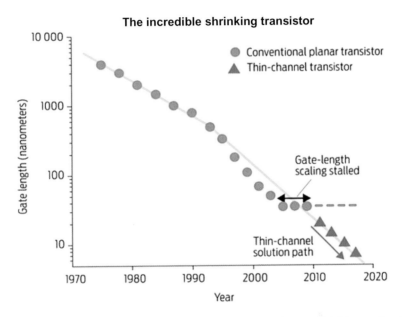

Figure 10. The increase in the number of transistors per chip has been enabled by continual reduction in transistor size (*Image Courtesy*: Khaled Ahmed, Intel).

shows that between 2004–2010, very little happened as a fundamental limit to manufacturing was reached, and then in 2011 a novel transistor architecture known as the thin-channel transistor was introduced, enabling the continued shrinkage.

Beyond that, there are fundamental limits to fabrication and manufacturing tolerances that cannot be overcome, so greater performance will be achieved by having multiple-core machines and perhaps via completely redesigning the architecture of chips. In the early days of computers, there was a direct and very strong correlation between the computer performance (how many calculations it could perform per second) and the number of transistors/valves present. As computer design has evolved, and there is more use of memory than ever before, the correlation is now weaker. Doubling the number of transistors no longer doubles the performance, but it does still cause it to increase. One thing to bear in mind is that at current rates, following the graph in Figure 10, by 2026, there should be around 100 billion transistors per microchip — comparable to the number of neurons in the human brain. With the recent developments in artificial intelligence, we have exciting times ahead within the next 10 years. Echoing Feynman's original dream of the ultimate in miniaturization, a transistor made using a single C_{60} molecule was demonstrated in 2000, although this was later shown to have serious issues with reproducibility, so never went beyond a fun and intriguing lab experiment.

Beyond Moore

All of these developments are being driven by economic forces and the desire of semiconductor companies to stay in existence. The past decade has seen developments in alternative technologies that will enable us to circumvent or harness some of the limitations due to be encountered by semiconductor devices such as quantum phenomena. This endeavor is known as "beyond Moore". This has paved the way for quantum technologies which are now in the ascendancy — creating chips, known as *quantum computers* (Figure 11), whose operation principle is based on the laws of quantum mechanics, which as we saw in Chapter 2, are pretty funky. These computers, considered until recently to be nothing more than a curiosity and probably a waste of time, have demonstrated that they can indeed carry out certain types of data processing far faster than conventional computers, and are of particular interest in cryptography and data sorting algorithms, and much of the initial impetus for their development came from the US military and the EU. In the late

Figure 11.   1000-bit quantum computer chip (*Image Courtesy*: D-Wave Systems Inc.).

90s and early 2000s, both funded quantum programs totalling several $100M, and just late in 2016, the EU embarked on funding €1 billion to turn lab experiments into commercial reality.

The quantum computer chip shown above was released on the market in 2015, containing 1,000 quantum bits, or "qubits", and has been put into use by Lockheed Martin, NASA, and Google among others. A more recent version, released in early 2017, contains 2018 qubits, and is used by Google, NASA, USRA, and defense companies. Quantum computers are becoming big business, although at the time of my writing this (summer 2018), there are ongoing controversies as to whether the chips are actually acting as quantum computers or not.

This slide toward quantum behavior has been inevitable and when the size of the features in a chip drops below around 5 nm, the effects that we mentioned in Chapter 2 such as quantum tunneling and interference will start to become apparent and will strongly influence the electrical behavior. Another issue to think about is the fact that when we get to the level of around 5 nm in size that is approximately 20 atoms across, so a cube $5 \times 5 \times 5$ nm contains $20 \times 20 \times 20 = 8,000$ atoms. Of those atoms, $20 \times 20 \times 6 = 2,400$ atoms, or just under a third, are on the surface, the remaining 5,600 atoms being inside, in the bulk. As surface atoms tend to have higher energy than bulk atoms, this particle will not behave like a bulk amount of the same material, either electrically or in terms of its color, or other properties as discussed throughout this book. This is something we notice with transistors in computer chips, that the random differences and defects that naturally occur in materials, which tend to be found more on surfaces than anywhere else, become apparent. What I am getting at is part of the larger problem which hampers the commercialization of some nanotechnology-based or inspired materials and devices — their inherent un-manufacturability. This sounds a bit grim and refers to the fact that for certain physical properties of materials, if particle size varies by more than a certain amount, this can have a detrimental influence on their operation in practice. As a case in point, we saw in Chapter 1 that the color of gold nanoparticles depends very sensitively on their size. If we wish to manufacture a pigment which is a particular color, and we find that just a few nm of variation in the size leads to a different color, then that would mean we would need to be able to guarantee that the process we use to manufacture the particles can do so with say, no more than 1 nm variation. At these lengthscales, this is incredibly demanding, and it just keeps getting harder to maintain sufficient precision as we make ever smaller things. It is no accident that the areas where nanotechnology has entered mainstream manufacturing involve the use of nanomaterials that are typically closer to 100 nm in size than 5 nm, and the applications are usually based on mechanical rather than electrical properties of these nanomaterials.

Where is Nanotechnology Actually Used?

At this point, it is worth having an overview of the electronic technologies that make use of nanotechnology in one way or another. We can broadly classify these into systems and circuit elements. By systems I mean data storage devices such as solid-state memory, hard disk drives, CDs, DVDs, and Blu-Ray, and by circuit elements I mean the individual components that are used to make these devices, such as transistors, which we have already covered. It is easiest to learn by example, so let us consider a smartphone. At its heart, it has digital circuits that store, access, and process our data, in the form of images, videos, songs, games, etc. This information is stored in a solid-state memory of the sort mentioned earlier, with transistors and capacitors. The information is usually accessed via a touchscreen user interface, which is a pressure-sensitive device. Different manufacturers use different physical phenomena to detect the pressure on the screen due to a finger, but the most common, and also the most energy-efficient is to detect the change in capacitance created when we push on a screen. This is measured using, you've guessed, a capacitor and a transistor. The screen itself has pixels, which as we discussed in Chapter 2 are typically around 50–100 microns across, so are in fact quite enormous in terms of nanometers, although they are made up of layers of materials which are often less than 100 nm thick. The screen is covered by a layer of glass, typically so-called "gorilla glass", developed by Corning, which is around 0.6 mm thick.

Gorilla glass (Figure 12) has a history spanning over 40 years, but only became the global brand recognized today when Apple asked Corning back in

Figure 12.   Infomercial for Gorilla glass for touchscreen devices (*Image Courtesy*: Corning Gorilla Glass).

Figure 13.   The first iPhone — the first application of Gorilla glass.

2005 if they had any glass which was strong yet thin and with increased resistance to cracking over and above normal glass. The story goes that the original prototype of the iPhone had a plastic screen, and that Steve Jobs tested it for a few weeks, including carrying it around in his pocket, along with his keys. He noticed that the keys scratched the screen and so, with barely a month to go before the product was due to be released, he demanded that the plastic screen be replaced with unscratchable glass (Figure 13).

Apple contacted Corning, the American glass manufacturer who had developed such a glass but had not considered it for the computer industry, and the rest, as they say, is history. The roughness of gorilla glass is well below 1 nm, so it is a super-smooth surface, an important aspect given that it is designed to be touched. What has any of this to do with nanotechnology? It is mostly that the fabrication techniques developed by nanotechnologists have enabled the transistors and capacitors to be created as small and powerful as they are and have helped to create brighter light-emitting materials for use in the pixels. The tools that we introduced in Chapter 4, particularly the atomic force microscope (AFM) is routinely used for quality control purposes to check the surface roughness of the glass and ensure it meets requirements. I should also point out that there is a version of Gorilla Glass that contains nanosized silver ions for their antibacterial and antimicrobial properties, branded as "Antimicrobial Corning Gorilla Glass".

What has nanotechnology to offer the semiconductor industry? There are vast research programs run across the EU, US, Asia, and Australia looking at this matter, and trying to support the existing industry as well as looking to the future.

Companies like Intel have known for many years that they need to look and plan beyond traditional devices and circuit architectures to ensure their future, and this is collectively known as "beyond Moore". This is a two-step process — (i) help make incremental improvements to existing technologies and processes and develop the next generation and next–next generation transistors and fabrication processes, and at the same time (ii) explore alternative materials and device architectures that may be commercially relevant in 20–30 years time. This involves looking at, for example, graphene devices, molecular electronics, novel materials, and quantum devices. This is a very fertile research area, the idea being to see if we can do what we currently do with silicon transistors and integrated circuits but using alternative materials in an attempt to circumvent the issues due to the inherent unmanufacturability of very small structures. Molecular electronics is a field that has been around, at least as an idea for over 50 years, the concept being that one way to overcome the naturally occurring statistical variations in properties of individual transistors is to make them all exactly the same. This is currently not possible due to the way in which we create them, by a process known as doping, a word which is in common use now with the Olympics and drugs scandals. The doping I am referring to is however a completely different use of the word. This is a process whereby we expose a material such as silicon to a hot gas of another material, typically phosphorous or boron. Atoms of these materials diffuse into the silicon and affect the number of electrons in it and where they are. If we do this in a controlled way, we can create different regions with different electrical properties and hence create a transistor. The problem is that we cannot control exactly *where* each atom goes, so there is an element of randomness involved. One way to control where each atom goes is to grow the transistor atom by atom, or in other words, use a molecule! The idea makes logical sense, and the only problem is to make electrical connections to these molecules, which is the sticking point. It turns out that for most molecules, the way in which they carry electrical current is via quantum tunneling, which gets exponentially less efficient as we try to move current over larger distances. Unfortunately, it only works if the molecules are no longer than around 1 nm. That is a bit small to be making electrical contacts onto, so a variety of clever ideas have been used over the years to do so, with varying levels of success. In each case, the nature of that electrical contact between the external wires and the molecules has proved to be more variable than the silicon transistors they were due to replace, so molecular electronic devices have yet to make any impact commercially.

This must not be taken out of context though, as one could argue that molecular electronics is already in common use, in OLED (organic light emitting diode) displays. These are rapidly replacing LED displays, as they emit sufficient light that they do not need a backlight and as a result can be used to make thinner displays than LEDs. Much research has been done on trying to improve this but with limited success. Many very interesting and ingenious papers have been published on the topic, and it is something that I personally have done some work on, with the emphasis being on reproducibility, which from my perspective is a problem and therefore a barrier to commercial implementation. Making electronic circuits at the nanometer scale is extraordinarily difficult and does suffer from issues with reproducibility — this is what led to the demise of carbon nanotube devices — they simply were not able to improve on silicon devices reliably. There is much hype and hope about graphene that it can do better (Figure 14). The jury will have to remain out on that one, but I again believe that although literally billions have been spent worldwide on this material and research into its properties, it is unlikely to make the impact that was touted at the beginning, at least not in the areas that were suggested.

Call me a sceptic, but this does seem like history repeating itself. There is no doubt that graphene, in principle, has truly impressive properties, both electrical and mechanical. However, it has the annoying habit of reacting to its environment, in that it attracts contaminants from the air, which stick on to it and modify its electrical properties. One way around this is to passivate it, which means encapsulating it in a different material so that it cannot interact in that way anymore.

The best devices made using graphene are free-standing, i.e. it is suspended over a hole in a material such as silicon dioxide. This way it cannot interact with the underlying surface, as there is none! As a result, its electrical properties are those of pure graphene, and some really rather impressive devices have been made

Figure 14.   Concept of a graphene transistor (*Image Courtesy*: Manchester University).

in this way. In Chapter 3, we touched on nanomaterials and composites incorporating graphene, which have found application in a range of areas due to potential improvements of the strength-to-weight ratio of materials. Another area where graphene is being heavily researched today is in optics — it turns out that as charge can flow so quickly in graphene, it is a suitable material for use in high speed lasers for telecommunications. As recently as 2016, a tuneable THz (terahertz) laser was demonstrated, based on graphene. There is a desire for such light sources in applications such as remote sensing and all-body scanners at airports. Whether graphene is the way forward will ultimately be decided on cost versus performance, as with just about everything else.

An example of the true application of nanotechnology to electronics is the memristor. This is a new electronic component created in 2008 by a team of researchers at HP labs in Palo Alto, and is essentially a resistor with memory. What this means is that it is a resistor whose resistance depends on what has happened to it in the past, i.e. how much voltage has been applied and when, so it is potentially an important data storage device. It was originally proposed by Leon Chua from Berkeley, who predicted that there would be a component with its properties. His prediction came about from looking at the properties of the equations used to describe electrical circuits, where the passive components are resistors, capacitors, and inductors. He noticed that the equations suggested the existence of a 4th component with the above properties and he called it the memristor. Similar things have happened in the past, when equations for a system suggest the existence of something hitherto undiscovered, the main examples being Dirac's quantum wave equation which predicted the existence of anti-matter, the equations in the standard model of particle physics which Higgs saw predicted the existence of the so-called Higg's boson, and Einstein's equations of general relativity which predicted the existence of gravitational waves. I am not saying that the memristor is as significant a finding as these other examples, but it is yet another example of when nature is kind to us and our equations to describe it actually work! Memristors are in principle simple to create and consist of a layer of oxide (usually titanium dioxide) sandwiched between two conductors. It seems that the passage of a current through the device causes oxygen atoms to move, affecting the resistance, which remains at this new value when the current is switched off, and this is how the memory works. Devices with this property of retaining information when they are powered off are said to be *non-volatile*. The initial devices had a footprint of 45 μm $\times$ 45 μm, but this has since shrunk to less than 50 nm $\times$ 50 nm, resulting in a comparable data density to that found in flash storage devices. While it indeed sounds extremely promising as a new technology, it would appear that the issues to do with the manufacturing cost and reliability have led to it being shelved, at least for now. In summer 2014, HP announced their latest breakthrough in computing and called it "The Machine", which was due to be produced by the end of the decade. This was meant to be by far the most powerful computer ever built, resulting from a complete redesign of the computer architecture, and would rely heavily on memristor memory. The way in which a conventional computer works is that information is stored in the memory, and is retrieved for processing, which has an overhead in time terms. The concept behind the machine is that the processing will be done inside the memory chip, thus potentially speeding up the computations by orders of magnitude. By the summer of 2015, the picture had changed and memristors were no longer on the table, instead replaced by more conventional memory. Little is understood about the exact mechanism by which the resistance of a memristor changes, and as long as that persists, it has little hope of long term success. An alternative which is similar but involving more layers has been independently developed by Intel and Micron, called the "3D XPoint memory", which entered the mass market in the latter half of 2017.

The development of computer memory is illustrated nicely in the below chart from Intel (Figure 15). Almost all conventional memory chips are based on NAND logic gates, which has been a winning technology for over 25 years, but as the trend toward miniaturization continues, there are inevitable stumbling blocks.

The largest of these is the fact that below around 25 nm, there are issues with repeatability, in particular, the number of read/write cycles that can be performed essentially drops by 50% for every 10 nm reduction in size (Figure 16). The sort of storage devices we use today can be rewritten up to 10,000 times but by the time they

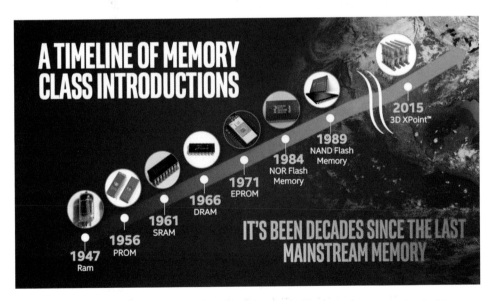

Figure 15. The development of solid-state computer memory, showing little has changed in terms of architecture since 1989 (*Source*: Intel).

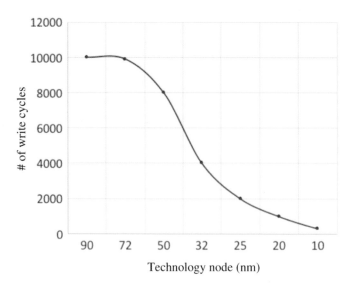

Figure 16.   Number of read/write cycles of NAND Flash memory devices as a function of the technology node.

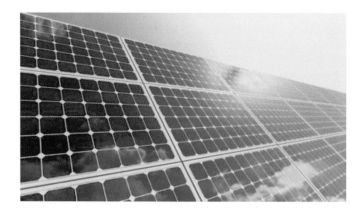

Figure 17.   The latest developments in low-cost, light and durable solar panels have been made possible by nanotechnology.

get to 10 nm across, this drops to a few hundred times at best. This appears to be a fundamental limit, which led to the realization in the mid-2000s that alternative device structures and materials would be needed if the miniaturization trend were to continue. On this basis, HP labs threw a significant proportion of their resources into the memristor work, but of course all that did was lay the foundations for others to improve on.

Another area where nanotechnology has made great strides, at least in the lab, is in solar cells, or photovoltaic (PV) panels (Figure 17). With the drive toward renewable or at least "green" energy, it is becoming more and more cost-effective to have solar panels incorporated in new-build houses. Our house for example has four such panels, which together are capable of producing around 1200 W of power on a sunny day, sufficient to run the washing machine on a low temperature cycle and the radio and a few other things, but not quite enough to power the entire house. These panels are typically made from monocrystalline silicon, which has a theoretical efficiency of around 33%, but in reality is usually around 11–12% for most household panels, rising up to 25% for high-end cells.

Such silicon PVs have been around for over 50 years and are coming down in price all the time. One of the issues with them is their weight — a typical domestic panel weighs around 10–20 kg/m^2, and costs around £130/m^2 to manufacture. Attempts have been made over the past 10 or so years to use nanotechnology to increase the efficiency of PVs, using cheap nanomaterials which would be light and durable. One of the core concepts here is to increase the surface area over which light is absorbed without having to make the PV too large — which can be achieved through having 3D structures, usually pillars, inside the cell. Such structures can be created using nanotechnology, and to date, the best efficiency recorded using a research-grade solar cell is just under 48%. As ever, there is a trade-off between efficiency, reliability, complexity, and cost, which is why silicon is still the most common commercial solar cell material. The industry-leading alternative is the so-called dye-sensitized solar cell which uses organic molecules to absorb light rather than silicon. These molecules have a lower cost than Silicon, can be mass-produced using solution chemistry, and the surface area for absorption of light is maximized by using titanium dioxide nanoparticles with diameters in the range 10–60 nm as a scaffold upon which the molecules that are used to absorb light are placed. This is a particularly interesting material as it is used in paints, sunscreen, cosmetics, and as a food coloring, and we have already encountered it in the memristor. The efficiency of dye-sensitized solar cells is typically around 12%, comparable to basic silicon cells, but their manufacturing cost has been estimated as around £40/m^2, less than a third of the cost for silicon, with the added advantage that they also weigh less. It has been forecast by the European Union Photovoltaic Roadmap that these dye-sensitized solar cells will be able to significantly contribute to renewable electricity generation by 2020. This is a direct example of the large-scale application of nanotechnology for something that is a benefit to society and our environment.

We have focused on electronic devices and components so far, but there is another class of device, which may well be where nanotechnology has a lot more to offer, which is electromechanical devices. These are microscopic devices that have moving parts, and where the motion can be measured using electrical means, or in other words, things whose electrical properties change when they move somehow. These are typically less than 1 mm in size, with characteristic dimensions on the micrometer scale. There is a lot to be gained through such

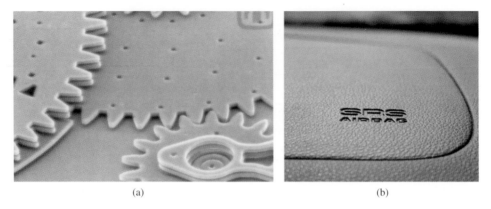

(a)   (b)

Figure 18.   (a) MEMS gear mechanism. The image size is around 80 microns, so the features are formed with a precision of better than 1 micron; (b) Airbags are triggered by MEMS sensors which can detect a collision.

miniaturization of structures in increased sensitivity, reduced power consumption, and ultimately reduced manufacturing cost over larger structures made using more conventional machining processes. We are already familiar with such devices, they are broadly given the name MEMS, or microelectromechanical systems (Figure 18). The processes used to create these structures is known as *micromachining*, and uses lithography to create tiny structures. In our daily lives, we encounter several of these in a variety of situations. In the event of a car crash, airbags are deployed. The sensors that detect a crash has occurred are called accelerometers, which literally detect when the car has undergone a significant acceleration, and these were the first large scale implementation of a MEMS device.

The second area where we unwittingly rely on MEMS devices is smartphones — they typically have several of these to detect the phone tilt and motion, and they are also used for optical image stabilization to move the lens in such a way as to cancel out the effect of a shaky hand.

A variety of other structures have been created using MEMS technology, including micro-engines, mini-drones, print heads for inkjet printers, gyroscopes for aircraft, microphones for mobile devices, car tyre pressure sensors, AFM cantilevers such as the ones we encountered in Chapter 4, and a number of others, and the market is growing all the time.

The first MEMS devices were created toward the end of the 1980s, and as the characteristic size of these devices has continued to shrink, they have started to enter the nanometer regime, and over the past few years, we have started seeing Nanoelectromechanical Systems, or NEMS devices appearing. A good rule of thumb is that the smaller something is, the more sensitive it is. Most MEMS/NEMS

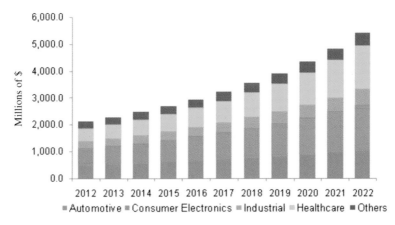

Figure 19.   The size of different global market sectors in $ Million for MEMS up to today and projected forward over the next 5 years (*Source*: Grand View Research Market report December 2016).

structures sense changes to their environment via changes to their mechanical properties, which manifest themselves as changes to their resonant frequencies. This is something that is actually rather straightforward to measure using simple electrical circuits. As with many things nano, manufacturing tolerances are an issue, meaning it is difficult to make the leap from the lab to the factory. A recent market report from Markets and Markets estimate the global NEMS market to reach $108.8 Million by 2022, which is around 40 times lower than that of MEMS (Figure 19). This is partly due to the glacial speed with which new technologies are adopted commercially and also to the fact that these smaller devices will have more niche applications. Arguably, the application space with the greatest impact and potential for these devices is in healthcare sensing, as we will explore in the next chapter.

Chapter 7

Nanotechnology in Healthcare

The doctor of the future will give no medicine, but will interest his patients in the care of the human frame, in a proper diet and in the cause and prevention of disease

— Thomas Edison

A Ridiculously Brief Introduction to Medicine

At the risk of spectacularly oversimplifying it, medicine is essentially the art and science of maintaining and prolonging our health and wellbeing (Figure 1) over and above what we can do for ourselves. This is done by using a multitude of approaches ranging from drugs, surgery, therapy — both physical and mental. In different parts of the world, the exact balance of each of these varies significantly depending on the GDP, societal expectations, whether the predominant philosophy is western or eastern, and other factors (Figure 2).

In a nutshell, eastern medicine is considered "natural" as it is based on the ability of the body to heal itself, so focuses on how diet can be altered using natural ingredients to optimize this for any given ailment, whereas western medicine is considered "unnatural" as it focuses on the disease itself and what therapies can be used to treat it. Each system has its successes and failings and far be it from me to even attempt to make any judgement on either. Nonetheless, it must be said that the body's own defenses are capable of dealing with a whole host of nasties, usually by producing antibodies to them. Antibodies are molecules that selectively attach to pathogens and destroy them, so at least from a philosophical way the eastern approach makes a lot of sense. If it works, that is…

Figure 1.   The rod of Asclepius — the symbol associated with medicine and healing, from Greek mythology. Often confused with Hermes' staff, the Caduceus.

(a)   (b)

Figure 2.   (a) Eastern, or Chinese medicine is largely herbal with the philosophy that under the right conditions, the body can heal itself — much along the lines of what Edison believed; (b) Antibodies selectively attach to pathogens and ultimately destroy them.

Medicine has come a long way, especially in the last 50 years or so, but at times, it is still rather barbaric to have to cut people open and perform invasive surgery to cut out or repair diseased parts. The challenges facing medicine have grown as people are now living longer than ever before, and our bodies are reaching their natural limits which we have to try to find ways to circumvent or reverse. Mass production of foodstuffs involving processing to enhance profit margins, food lifetime, taste, ease of preparation, etc., have unintentionally led to the proliferation of many diseases, compounded by the use of pesticides in agriculture and hormones in livestock feed. The balance of fat, sugar, protein, and fiber in many foods has shifted to unhealthy levels over the past 2–3 decades which doesn't help either and has contributed to the rise in obesity. According to a recent study published by the Office for National Statistics (ONS) in the UK, over half of men and women in the UK are classed as overweight or obese. To be fair, overweight means having a body mass index (BMI), of 25–30, and obese is a BMI of over 30. This is rather arbitrary, as it does not take into account above-average muscle mass, so most body-builders would be classed as at least overweight. The BMI is measured as the weight in kg divided by the square of the person's height in meters. It is known that it is the fat in the abdominal region that carries the greatest health risk, so work is ongoing to directly measure that in order that preventive steps can be taken to avoid issues with heart disease, type-2 diabetes, and hypertension. Diet and appropriate exercise are well known to be the levers we as individuals can control in order to obtain and maintain ideal body weight and health. Our busy lives with their associated stresses and socio-economic factors however mean that while we all know what we should and should not eat etc., we don't always get it right, and tend to eat and drink more than we should, and exercise less than we should. Edison did have a point when he made the statement I have reproduced at the beginning of this chapter, but of course it is just not that simple. Genetics also plays a significant role in determining our risk factors, even if we do everything as we should. Nonetheless, that doesn't mean we shouldn't try!

We should eat more of this to maintain a healthy weight and amount of body fat.

And less of this…

I'm sure this fellow agrees…

On the whole, our society sadly tends to place more emphasis on the value of food than its quality. Edison's idea was pragmatic in the extreme — if we all took care of our own health then we could avoid illness, and if we don't take care, any disease is our own fault. We must remember that his statement came at a time before we truly understood the effect of genetics. It has come to light in recent months that many people in, particularly the western world, have inherited a significant proportion of genes of Neanderthal origin, which unfortunately predisposes us to many diseases. It has been long known that certain diseases, including heart disease, breast cancer, and Parkinson's do have a genetic link. I remind myself of this when I drink that glass of wine on a Friday evening, that genetically at least, I am not predisposed to oral cancer, alcoholism, or pretty much anything for that matter (Figure 3). Fingers crossed. I'll drink to that. Any excuse!

Advances in medical treatments over the past 80 years have enabled lifespan to increase by almost 20 years as shown in the figure demonstrating life expectancy as a function of birth year in the UK (Figure 4).

Figure 3.   It's purely medicinal, honest Doc!

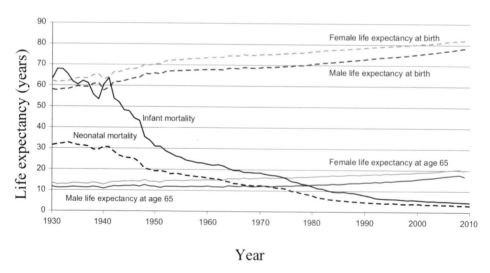

Figure 4.   Life expectancy and infant mortality as a function of birth year (Data from the Office of National Statistics, UK).

This shows that there has been a steady increase in life expectancy, of 20 years over the past 80 years, with the most dramatic improvement being in the decrease of infant mortality dropping by almost a factor of 20 over the same period. The primary factors leading to infant death in the UK are infection and asphyxia during birth. In the early 1940s, penicillin started to become widely available, leading to the sudden drop in mortality from 1945–1950. Advances in in-utero diagnostics have contributed to the continued decrease since then. On the whole, looking at this data, it is clear that there is a steady trend regarding life expectancy (apart from a blip during the Second World War when life expectancy decreased a bit), but there have been no dramatic changes — at no point has a discovery been made which has altered this trend overnight. The elixir of life has not been discovered, and instead we know, as did our ancestors, that on average, if you want to live longer then you must look after yourself. Unfortunately, there is no pill we can take that increases lifespan!

In order to see where the greatest need for intervention is, we need to understand what the key areas in medicine are. If we look at the reasons why people visit their GP or walk-in health centers, the results of a survey carried out by the Accent group on behalf of the UK government in 2013 are that the most common ones are (i) flu-like symptoms such as coughs, colds, and sore throats (8%) (Figure 5(a)); (ii) various skin conditions (8%); (iii) ear/eye infections (7%); (iv) sprains/strains (6%); (v) blood-pressure check/other health check (6%) and (vi) cuts, bruises and abrasions (5%). None of these are areas where nanotechnology can make a direct impact, although as we have seen in the introduction, silver-impregnated plasters are a useful antiseptic treatment for cuts and abrasions.

If this is the case then, what does nanotechnology have to offer in the field of medicine? The tools that we have been discussing for the past few chapters

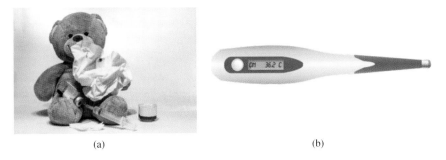

(a)   (b)

Figure 5.   (a) Flu-like symptoms are the most common reason for visits to the GP; (b) Nanotechnology has most to offer in the detection and treatment of disease.

Figure 6.   What nanomedicine is NOT — nanorobots going around repairing damaged cells!!!

find ready application in the diagnosis and treatment of more acute illnesses. There are two broad areas where nanotechnology has started to be transformative:

(i) in the development of more sensitive sensors for the early detection of the chemical markers of disease, and (ii) in therapeutics — targeted delivery of medicines especially in the case of chemotherapy drugs for cancer treatment (Figure 5(b)). As mentioned at the beginning of this book, nanotechnology is NOT about making nanorobots which can autonomously go around our bodies repairing damaged cells (Figure 6). So, the image above is pretty, but total twaddle! However, as we will see later, while we cannot make robots that repair cells, we *can* make nanoparticles that attach to and destroy cancer cells.

Cancer — The Basics

Cancer is one of those diseases that touches us all, either directly or indirectly. In the UK, the lifetime risk, i.e. the risk of being diagnosed with some form of cancer at some point during our lifetime in those born since 1960 is at 50%, as compared to 39% in the United States. Survival rates are most commonly expressed as the percentage of those who are still alive 5 years after diagnosis, and this figure has dramatically improved over the last 40 years for all forms of cancer, most notably for stomach cancer. This data, obtained from the ONS, is reproduced on the next page. We can see that an area where more research is desperately needed is

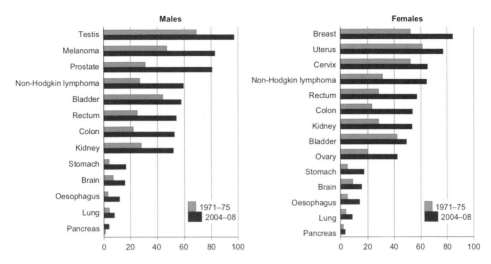

Figure 7.   5-year survival rates for common cancers in the UK (Data from ONS 1995, ONS 2010a, London School of Hygiene and Tropical Medicine).

pancreatic cancer. Although the survival rates have increased significantly, they are still unacceptably low (Figure 7).

The survival rate for prostate cancer has also increased dramatically in this time period. There are a multitude of reasons for these improvements across the board including better diagnostic tests that are capable of detecting cancer earlier and an ever-increasing arsenal of drugs and therapies capable of targeting specific cancers.

Significant amounts of funding have gone into cancer research and at any one time, multiple approaches ranging from gene therapy to chemotherapy and radio-therapy are being investigated for all of the main forms of cancer. Most hospitals now also have advanced imaging capability, especially MRI (magnetic resonance imaging) scanners that can detect cancer tumors with millimeter-scale resolution, so we have a better idea of where exactly it is and how it is spreading (Figure 8). Still, I think it is fair to say that once cancer starts, it is essentially a race between it and the clinician/patient to try to eliminate it before it is too late. We are not yet at a point where we can prevent it from ever starting, which is undoubtedly what we should be aiming for.

In line with the other topics covered in this book, we will look at the fundamen-tals of the problem, in as much as I understand them at least, and then look at the steps that are being taken or indeed should be taken by nanotechnologists together with clinicians to try to tackle this problem. I have chosen cancer in particular sim-ply due to its sheer impact on us. We will start by looking at what cancer is, then how it is/may be detected, followed by a look at the detection schemes that are being

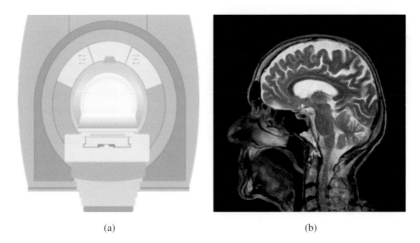

(a)   (b)

Figure 8.   (a) MRI scanners allow us to image inside the body and are a powerful way of finding tumors; (b) An MRI image of a human head showing the brain in great detail.

developed. We will then look at the limitations of current therapies and see what nanotechnology has done to help. This is a very exciting field with many parallel developments that are happening on a daily basis, so I am aware that by the time this book comes out, there will have been breakthroughs that I have not talked about.

The first question we should address is "what is cancer?" The world health organization (WHO) has a useful definition which reads as follows: *cancer is a generic term for a large group of diseases characterized by the growth of abnormal cells beyond their usual boundaries that can then invade adjoining parts of the body and/or spread to other organs.* By "abnormal" cells, they mean cells that reproduce faster than usual, hence their spreading beyond their usual boundaries. This creates the primary tumor over time. The spreading of cancer cells from one part of the body to another, known as metastasis (Figure 9(a)), gives rise to secondaries, which are often deeply problematic as they can spread rapidly through the bloodstream or through the lymphatic system.

Cancer can start as a single cell that for some reason has developed a fault. Such faults are usually triggered by a mutation in the DNA inside the cell (Figure 9(b)), which can happen for all manner of reasons including: (i) random mutations when the cell was formed, (ii) spontaneous mutation after the cell was formed, (iii) exposure to a foreign object/substance which persists for an extended period, such as soot or asbestos in the case of lung cancer, (iv) exposure to UV sunlight which directly mutates DNA in the case of skin cancers, (v) exposure to certain toxins or bacteria, as appears to be the case in some forms of stomach cancer. A very useful graphic illustrating the differences between normal and cancer cells is reproduced in Figures 10 and 11. Cancer cells tend to have multiple nuclei that take up much of the volume of the cell, and have irregular shapes as shown.

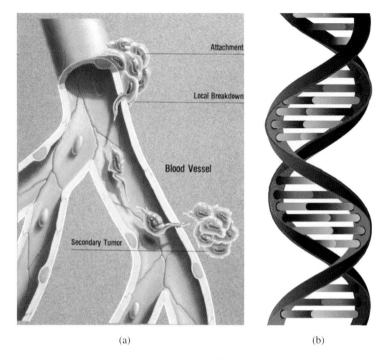

(a)   (b)

Figure 9.   (a) Metastases — the spreading of cancer, usually through blood or lymph vessels; (b) Most cancers are triggered by mutations in our DNA, and understanding the causes of this will aid in prevention of the disease.

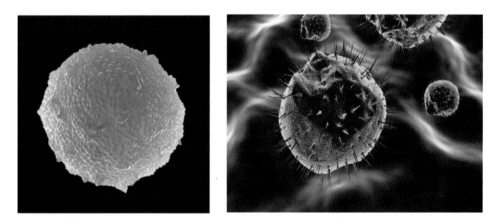

Figure 10.   Normal versus Cancer Cell. On the left is a normal cell, in the case a white blood cell, whereas on the right is a cancer cell. Note the spiky, rough surface.

Research has been carried out on cancer cells for decades since before the mid-20th century and we have learned much about their structure, what makes them different to normal cells, and are starting to unlock how to destroy them. Much of the core research over the last 60 or so years has been carried out on so-called HeLa cells.

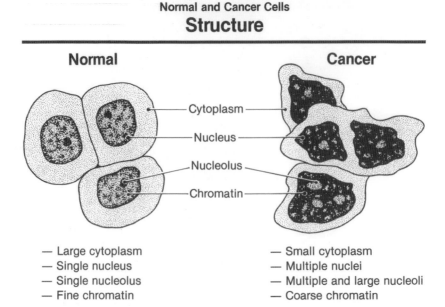

Normal and Cancer Cells
Structure

Normal	Cancer

- Cytoplasm
- Nucleus
- Nucleolus
- Chromatin

Normal	Cancer
— Large cytoplasm	— Small cytoplasm
— Single nucleus	— Multiple nuclei
— Single nucleolus	— Multiple and large nucleoli
— Fine chromatin	— Coarse chromatin

Figure 11.   The differences between normal and cancer cells. Image from Pat Kenny, NIH.

Sounds like some sort of Latin or Greek name, or a Marvel character (there is a character with that name who is a foe of Thor). As it happens, it is much more interesting. They are named after their source — Henrietta Lacks, who was an African–American woman who died from cervical cancer in 1951. Her doctor at Johns Hopkins hospital performed a biopsy and diagnosed cancer. A cell biologist at the hospital, George Gey, noticed that these cells were able to reproduce unlike any other cancer cells discovered up until that point. This was a significant finding as much of cancer research until then was hampered by the difficulty in keeping cancer cells alive for long enough to carry out any experiments. The HeLa cells were so prolific that they could reproduce easily using simple cell culture media, and so Gey propagated the cell line and shared it with other researchers (Figure 12). These unusual cells, which were otherwise typical cancer cells, have become "immortalized" in that they are still reproducing and being used widely today. In fact, in common use are 6–7 such immortalized cell lines that have come from various individuals over the past few decades, and that are used by cancer researchers all over the world, as standardized samples for testing the efficacy of new therapies.

The typical presentation of a cancer patient is that they present at their GP with a set of non-specific complaints possibly including unexplained weight loss, fatigue, pain, shortness of breath, bleeding, changes in skin color, fever, and many others. In most cases, the signs and symptoms can be associated with many other

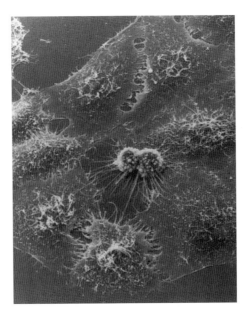

Figure 12.   Gey's first image of HeLa cells taken using an electron microscope, in 1951.

illnesses, so cancer is not necessarily the first thing on anyone's mind, unless there are known risk factors, a strong family history or an obvious lump. If symptoms persist, additional tests can be ordered and when the problem area is physically located, a biopsy can be taken. The most common diagnostic tool for cancer is to look at cells/tissue thus obtained in a microscope — cancer cells are easy to spot due to their appearance. At this point, if cancer cells are indeed found, MRI or a similar imaging tool is employed to determine the extent of the spread, so that an informed choice can be made about appropriate treatment. In many cases, the weak point is that very first step — either the patient noticing something not quite right and not doing anything about it soon enough, or a GP not recognizing the symptoms and keeping track of how the illness progresses with the patient. This all highlights the need for point-of-care diagnostic tools that are reliable, easy and cheap to use, so that when anyone goes to their GP, they can have a simple breath, blood, or other bodily fluid test which can yield immediate results, saving precious days or weeks. Such systems are called "lab on a chip" devices, and we will have a look at some of these shortly. What these detection schemes are based on is the fact that there are other ways in which cancer affects our bodies apart from those non-specific ones mentioned above. Cancer cells are essentially foreign bodies so our immune systems respond by producing antibodies to target the abnormal glycoproteins that cancer cells create. The particular antibodies are specific to each form of cancer and if detected, can indicate the presence of a tumor.

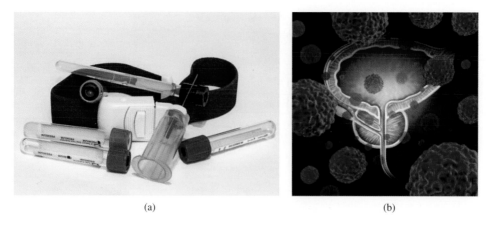

(a)   (b)

Figure 13.   (a) The paraphernalia required to take a blood sample for routine testing; (b) A number of diagnostic tools for prostate cancer have been developed by nanotechnologists.

Therefore, systems that can carry out simple blood screening to look for the presence of these antibodies can in principle detect several forms of cancer (Figure 13). The trick is to look for the right combination of antibodies, and to do it quickly, reliably, and cheaply. An example of where this is important is in the detection of prostate cancer. This is typically done by carrying out an assay for PSA (prostate specific antigen). If high levels of the antigen are found, further tests are carried out, which are usually invasive and rather unpleasant in order to determine if there are cancer cells in the prostate. The PSA test is notoriously unreliable as up to 60% of those showing an abnormal level of PSA do not have cancer, and around 30% of those with prostate cancer have normal PSA levels. Digging a little deeper, it turns out that there are several types of PSA, and the current test simply looks at the total amount (called the total prostate specific antigen, or tPSA). For the past 5 or so years, more specific tests have been developed that look at the amounts of the particular forms of PSA that are present, and clinical trials are ongoing to see if this can lead to more accurate diagnosis. We will see shortly that this approach of looking for multiple biomarkers simultaneously is where nanotechnology has something to offer.

Lab-on-a-chip — Small is Good

There are currently few such lab-on-a-chip tests in routine use (Figure 14(a)), despite early promise and vast sums being spent on this. It is still the norm that when we go to our GP with any ailment requiring a simple test, a sample needs to be taken and sent to a lab, with results typically taking up to a week to come back. With some diseases/conditions, this is too long. No doubt the most commonly

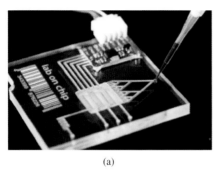

(a)   (b)

Figure 14.   (a) Lab-on-a-chip — A device that integrates sample handling and testing all in one go; (b) Using a blood glucose test meter — the first mass-produced lab-on-a-chip device. A small drop of blood is sucked, by capillary action, into the active area between electrodes.

used and the most reliable lab-on-a-chip device is the blood glucose meter test strip. The basic concept was devised by Allen Hill from Oxford University's Chemistry Department in the 1980s. While researching topics in electrochemistry, they recognized that when glucose gets broken down in the body using the enzyme glucose oxidase, there is a release of electrons. They came up with the idea of coating some metal electrodes with glucose oxidase, and then when a drop of blood is added, the glucose in it gets broken down, and the electrodes capture the electrons. This is measured as a small electric current between the electrodes, where the amount of current depends directly on the amount of glucose present. This in turn depends on (a) how much blood is added and (b) how much glucose is in that blood, which is what we actually want to measure (Figure 14(b)). Therefore, in order to be repeatable and reliable, the same amount of blood must be added each time the test is done. This is where microfluidics comes in — the test strip has a small channel etched into it with a specific geometry and size which is designed such that it collects and delivers precisely the right amount of blood to the electrodes. I guess that's why these strips sell for about 50p each as their fabrication is not totally trivial and they have to be made under highly controlled conditions. Attempts have been made to detect other substances in the blood using this kind of electrochemical approach, but so far there is little on the market.

At this point, I would like to point out a cautionary tale, but in a way that won't lead to my facing litigation… Perhaps the fastest rising star in the whole lab-on-a-chip universe was Theranos, a US-based company started by a Stanford Chemical Engineering Student, Elizabeth Holmes in 2003, rose to dizzying heights in terms of what they said they would be able to do using their revolutionary equipment which could take a miniscule drop of blood and analyze it for a whole range of diseases in real time, with clinics set up in a number of states (Figure 15).

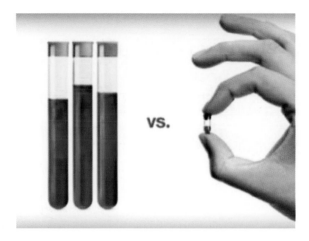

Figure 15.   Theranos was set up with the avowed aim of simplifying blood testing using highly sensitive equipment that would only require very small quantities of blood as compared to conventional testing (Image from Youtube/CNBC and Theranos).

The vision was that everybody could be screened for disease quickly, easily, and cheaply, without needing to go to hospital or even a GP. Unfortunately, it seems that the lab-on-a-chip technology they developed doesn't work very well, at least not yet. I am sure that given time, patience, fewer lawsuits and bad press and lots of cash and research, this vision may come to pass. Just not right now.

Let's Get Nosey

What other diagnostic schemes are there that have come about due to nanotechnology? There are two systems we should consider, one of which has been highly successful and is widely used, and one which has been under development for over 20 years with some individually spectacular results but has remained a research tool without wide application — yet. The latter is based on AFM (atomic force microscope) technology. The concept is as shown in Figure 16: an AFM cantilever is a highly sensitive mechanical structure that can measure tiny changes in temperature, stress, magnetic field, electric field, and mass. This technology was originally invented by Christoph Gerber, one of the inventors of the STM and the AFM.

The fantastic sensitivity is entirely due to its small size and the fact that a small change in any of the above conditions can lead to a measurable deflection of the cantilever. From a diagnostic perspective, the idea is simple — if we wish to detect a substance (which almost always comes in the form of a molecule) which is a marker for a specific disease, we functionalize (a fancy word we use for coating) a cantilever with another molecule that is known to react with the molecule of

Figure 16.   8-cantilever array used for multi-target sensing.

interest, which we will call the target molecule. As soon as we introduce the target molecule to the cantilever, either as a vapor (for the case of volatile compounds such as alcohol) or in a blood sample, then it will react with the detector molecule on the cantilever. It turns out that this leads to a bending of the cantilever due to a change in the stress on its surface, and this bending can be detected using an AFM setup (Figure 17). This concept was first demonstrated in 1994.

Unfortunately, there are a multitude of reasons why an AFM cantilever may bend, that have little to do with any chemical reactions occurring on them, such as changes in temperature, mass, liquid pressure, humidity, and other factors. To mitigate against this, the standard practice now is to use a reference cantilever, i.e. one with identical properties (geometry and size) to the sensing cantilever and measure the *difference* in their deflections. In this way, we can measure the actual deflection due to the molecular interaction of interest. This is a very sensitive detector and has been shown many times to be capable of detecting a variety of substances. Even better, early on, a cantilever array chip was produced by IBM (who were the pioneers in this technology and where Gerber was based) containing eight cantilevers, each with a different coating, which was used to detect the presence of several materials simultaneously. This was called the *nanomechanical nose*. Using suitable algorithms, it was possible to convert the deflection of each element in the array of cantilevers into a fingerprint unique to any given substance. In a move typical of the scientists involved, one of the first fields they applied this to was to see if they could discriminate between different types of whiskey (Figure 18)! The experiment was a simple one — they produced several of their array chips and measured how the cantilevers deflected when exposed to the vapor produced by a small volume of whiskey nearby. They saw that each type of whiskey they looked at gave a slightly different distribution of deflections of each cantilever, so produced a characteristic fingerprint for each one. A fun experiment no doubt, but potentially useful as a step in quality control.

Figure 17.   Illustration of a cantilever array, each of which has different molecules attached to it. When exposed to a sample, either gas or liquid, each cantilever will interact with components of that sample and bend accordingly. The combination of deflections can be used to detect minute amounts of material — particularly useful for multi-target sensing in early stage cancer diagnostics (*Image Courtesy*: Rachel McKendry & N. Kappeler, UCL).

Figure 18.   Yes — high technology capable of detecting minute quantities of volatile compounds was tested on…whiskey!

The work was expanded on by researchers at the London Centre for Nanotechnology (LCN) who applied it to measure alcohol content, density, and viscosity in a variety of beers. This was an extremely useful concept and in principle could easily be added into the brewing process — beer quality could be rapidly tested using tiny quantities. As is often the case though, the technique is not yet mature enough to be applied in a routine manner, so the cost involved in running these kinds of test are simply too high. Given sufficient impetus from any industry sector this could be turned around, but one thing for certain is that by and large, big industry is *very* conservative so is slow to take up new technologies. This is best left to small startups, which naturally take some time to become established. This is a key observation of course — just because something is technically possible and may even have been demonstrated in a lab, it can take decades to become deployed. There are many superb papers in journals such as *Nature* and *Science* where this array technology has been used for rapid sensing of various substances in blood, breath, and other bodily fluids. The arrays in most common use are the eight cantilever ones mentioned above, which are relatively straightforward to individually coat. In the field of data storage, arrays containing thousands of cantilevers have been produced, and the hope is that these chips could potentially be used for the above sensing purposes. The challenge there is that these arrays tend to be rather small, on the order of a few mm across, with the individual cantilevers only a few microns apart. This is simply too close to be able to differentially functionalize each cantilever using conventional methods, which are usually based on dipping or inkjet printing (which has a resolution typically of the order 10–20 microns), but developments are underway to work around this. Of course, another issue is to do with the cost — if these arrays are reusable then if they cost a few hundred dollars each to produce then that is workable, but if they are only single-use and need to be disposed of then that is a different story. Further work needs to be done on large-scale integration of the cantilever arrays with the sensing electronics and signal processing as well as fluid handling in order to be able to mass produce them cheaply enough. Why bother? There are more sensitive, reliable, and cheap techniques in use today, so what advantages does this technology have that makes researchers want to pursue it? The key advantages of this measurement modality are that it is non-invasive, label-free (i.e. we don't need to add any additional molecules which fluoresce or can be easily detected in order to see the molecules we are interested in) and requires only very small quantities of bodily fluid to work, i.e. tens of microliters compared to the milliliters required by alternative techniques. The label-free bit is crucial as it means that no additional lengthy or costly chemical treatment is needed — we can simply take a blood or serum sample and add it directly to the cantilever array which itself is usually immersed in a liquid such as water. Having multiple cantilevers means we can detect the presence of multiple biomarkers simultaneously, which is necessary

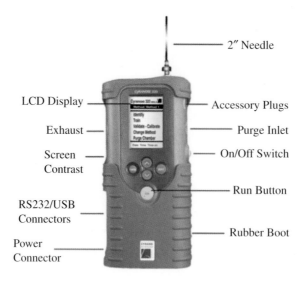

Figure 19.   The Cyranose 320 electronic nose (*Image Courtesy*: Sensigent).

as many diseases including cancer cannot be detected via the presence of a single target molecule. Just in the last 5 years, there are several cases, particularly in Switzerland, the US, and the UK where these cantilever arrays are being used to detect (i) volatile organic compounds (VOCs) in the breath, as a diagnostic for head and neck cancer; (ii) bedside monitoring of patients — these sensors are being used to detect genetic mutations in melanoma and breast cancer biopsies.

As for the other diagnostic sensor type, it has been by far the most successful and simple to implement. It is known as the *electronic nose* (Figure 19), and is purely electronic in its operation, as opposed to mechanical like the nanomechanical nose. At the heart of it is a transistor or an array of transistors, of the type used in digital circuits, known as MOSFETs, as we saw in Chapter 5. The principle of operation is very simple which is one of the reasons why it has been so readily adopted.

The idea is that the material being detected (almost always in the form of a gas consisting of molecules) is collected, ionized, and then passed over the transistor array, which the charged molecules then stick to. The electrical resistance of the transistors is changed when these charged molecules adsorb onto their surface, and this is very straightforward to detect. These devices can be used many times over without needing to replace the transistor array. In the example shown in Figure 19, the detector consists of an array of 32 transistors, which are coated with a nanocomposite material which collects (absorbs) the volatile compounds of interest. This is where the true value of nanotechnology has been brought to bear — each transistor has a different coating which is chemically sensitive to different types of molecules, therefore enabling rapid detection of complex mixtures. This has been applied to a wide range of problems and is currently being used in the US (the Cleveland Clinic) for lung and throat cancer detection via exhaled breath analysis. Detection of other forms of cancer is undergoing clinical trials and is showing great promise. The technology is there in principle to detect many diseases, it just needs link-ups between scientists and clinicians to move things forwards which seems to happen more by chance than by design. It has been "known" for a long time that the smell of one's breath can be used to indicate the presence of a variety of diseases, but this has not been quantifiably measured until very recently. I am of course referring to Chinese medicine, which uses the smell of bodily discharges of all sorts to indicate the health of specific organs. Apart from a few obvious cases such as ketoacidosis from diabetes, I find the concrete evidence of this to be rather lacking and it does seem to be purely anecdotal. More recently, it has been claimed that highly trained sniffer dogs can be used to detect melanomas just by smelling one's skin, and prostate cancer by sniffing urine. While there are good reasons to expect each form of cancer to be associated with biomarkers that will have a distinctive smell, the scientific evidence behind these claims is wooly at best. I'm afraid we are better off sticking with quantifiable, measurable effects for diagnosis, and perhaps leaving the dog's role to that of comforting companion rather than doctor!

What about other ways in which cancer can be detected? Other approaches rely on the fact that the chemical properties of cancer cells are different to those of normal cells, so we can attach things to them specifically, such as fluorescent molecules or other markers that we can easily see. We will briefly look into this now.

Nanoparticles — The Good Ones

We have touched on the application of silver nanoparticles for antibiotic treatment a number of times. As it turns out, they are also potentially useful as antivirals. It is known that colloidal silver particles (Figure 20), with a diameter well below 100 nm are highly effective at killing the human papilloma virus, or HPV. This virus triggers cervical cancer and is also to be found in warts. Several new over-the-counter wart treatments now take advantage of this. It has also been shown that the same nanoparticles may be used to destroy the AIDS and Hepatitis B viruses, by blocking their ability to replicate. This is incredibly important due to issues with the rise of drug-resistant strains of bacteria and viruses. The exact mechanism of action of silver nanoparticles is not fully known, but the evidence is mounting that it catalyzes redox (reduction and oxidation) reactions which ultimately lead to the destruction of such materials.

One of the most important and commonly-used tools in cancer research is the optical microscope. In a typical experiment, cancer cells are taken and fixed to a

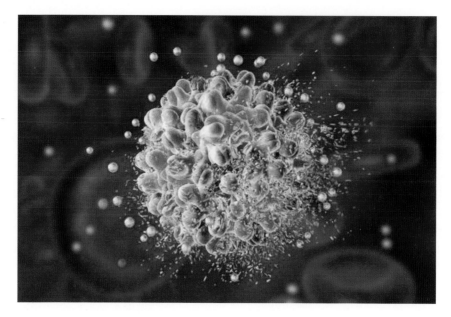

Figure 20.   Silver nanoparticles have been shown to be capable of destroying HPV, AIDS, and Hepatitis B viruses as well as many bacteria.

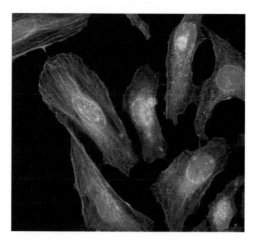

Figure 21.   Hela cancer cells as seen in a fluorescence microscope, showing DNA (blue), actin (green) and vimentin (red).

glass slide, then *stained* with a variety of chemicals to make specific parts of the cells stand out in terms of their color. A particular favorite of mine is shown in Figure 21 — this is an optical microscope image of Hela cells where actin, vimentin, and DNA are stained red, green, and blue, respectively. Actin and vimentin are proteins that create the scaffold (cytoskeleton) that give cells structure. The way in which these stains work is they absorb light from a lamp in the microscope, which causes them to fluoresce — i.e. to re-emit light of longer wavelength, i.e. a different color. Through the use of appropriate color filters, an image is generated using just the colors of the fluorescing stains.

The beauty of this is that the stains are specific — they attach to particular chemicals in the cell, so can be used to show the location of different cell components. This is not a diagnostic, it is purely a technique to help take clearer images of cells. The next step is to add some sort of diagnostic function to something which can attach specifically to cancer cells and then flag up that they are there. I am of course referring to the use of nanoparticles to be used as tags to attach to cancer cells via appropriate molecular aides, and then sense changes in the biological environment with a sensitivity hitherto unseen — i.e. at the molecular level. This has been demonstrated a number of times in different scenarios. Almost all of these approaches use the same concept — a metal nanoparticle, usually just a few nm in diameter and coated with *something* that makes it preferentially attach to the cell type of interest. This is known as a *core–shell* structure (Figure 22).

What that coating is determines what the nanoparticle can do, i.e. whether it is used for imaging (diagnosis), therapy, or drug delivery. Most approaches use gold as the nanoparticle material, and with diameter of order 2–3 nm. Gold is particularly stable, the nanoparticle size can be easily tuned, it is around the same size as many proteins, and it is a material not rejected by our bodies. The practicality is that the nanoparticles are prepared, coated with the protein/antibody/fluorescent molecule or drug of interest (this process is called *functionalization*), then injected into the bloodstream. When the particles encounter the cancer cells they have been targeted toward, they attach to them. They can then be detected using

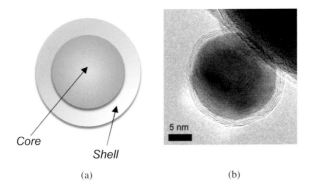

Core

Shell

(a)   (b)

Figure 22. (a) A core–shell nanoparticle — a core made from one material with a shell made from another, often for protection; (b) Metal nanoparticle (in this case, made from cobalt, with a graphene shell) (By Supermaster 2011 — Own work).

Figure 23.   An example of fluorescence — In this naturally occurring in the greeneye fish. *Top image*: as seen using visible light; *Bottom image*: as seen in UV light. Similar images are taken of test animals where the fluorescence is from nanoparticles attached to cancer cells.

various sensing techniques. This approach has not yet been clinically tested in humans, although it has been in human tissue samples and in living mice. We will have to leave discussions on the ethics of this for another day — this is something that all medical researchers face and I merely wish to consider what nanotechnology has to add. At the end of the day, new drugs/diagnostics and therapies need to be tested before they can undergo clinical trials and then be potentially used in humans. We don't have to like it, but if we wish to benefit from advances in medicine then for now, this seems to be the only way.

There are many imaging techniques that are capable of picking up fluorescence from within the body (Figure 23) — certainly from within mice as they are relatively small and can be bred with translucent skin, so tumors can be located and imaged without having to operate at all. Hopefully some day, this will be the case in humans.

To my mind, the most exciting advances that have happened over the past 20 years are the next step along from what I have just described — using these nanoparticles to destroy dangerous cells — like a magic bullet that seeks cancer cells and kills them. This brings me to a point that I feel needs to be clarified and has been compounded by unscrupulous advertising for certain painkillers. When we take a medicine, be it orally or intravenously, it enters the bloodstream and then goes everywhere in our bodies. It does not *target* specific areas, despite what any manufacturers may claim! If we have a headache and take a painkiller, it does not just all go to our head, gather there and tackle the pain. It literally goes everywhere. If medicines were targeted to specific problem areas, then we could take much smaller doses and greatly reduce the side-effects. For instance, ibuprofen interacts with the lining of the stomach as does aspirin, and paracetamol breaks down into toxic substances in our livers, so while they all help reduce pain/inflammation or temperature, they do have side-effects. This is even more pronounced in cancer treatment — chemotherapy, the side effects of which are profound, debilitating and deeply unpleasant. The ability to target the medicine to the affected area would alleviate much suffering, and this is an area where nanotechnology is expected to make great strides over the coming years.

In the field of medical diagnostics, iron nanoparticles (smaller than 20 nm in diameter) have been used to enhance contrast in MRI images since the mid-1990s, and it was demonstrated just a few years later in 1999 that in the presence of an alternating magnetic field (basically radio waves), these nanoparticles can absorb energy and heat up local fluids, and that this could be used to destroy cancer cells. Over the years, subtle improvements in specificity and efficiency were achieved, and then since the mid-2000s, work has been done on looking at alternative metal nanoparticles that are more controllable in terms of their properties, and that do not require a large, expensive MRI machine in order to operate.

The cornerstone of almost all nanoparticle-based medical research is that metal nanoparticles selectively absorb certain wavelengths of light or microwaves — tunable by controlling the particle size, as we saw in the first chapter. They can be created such that they absorb wavelengths that the body does not absorb. In a recent study, they were tailored to absorb infrared light, and then they were heated to the point where they were able to destroy local tissue. This approach has the advantages that there are no nasty drugs involved, but it is of course early days yet and it will not be applicable to all forms of cancer.

More recent developments have taken things a step further — the so-called magic bullet mentioned earlier. It is possible to use nanoparticles with multiple layers, or shells, each of which has a different purpose, starting with a protective coating on the outside, followed by a layer that allows attachment to a cancer cell type of interest, ultimately followed by a layer of "payload", or the medicine we wish to deliver to that cancer cell. There are a number of variations on this theme with different degrees of complexity, some of which even use the metal nanopar-ticle as a source of electrons that can trigger or enhance the operation of chemo-therapy drugs. This concept has been called a Trojan horse, as the cancer cells let the nanoparticles attach to them, unaware of their deadly intent. This is illustrated in Figures 24 and 25. So far this has been demonstrated to be highly effective in treating glioblastoma — brain tumors. As I write this, clinical trials are underway to test this in human patients.

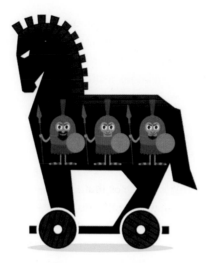

Figure 24.   Nanotechnology has pioneered the use of "Trojan-horse" particles to target cancer cells. They consist of particles, usually around 100 nm in diameter that contain a potent drug (the core) surrounded by a thin shell of an innocuous material that the body accepts.

Figure 25.   The ultimate magic bullet — a nanoparticle with successive layers, each of which has a different function, all aimed at delivering a drug payload to a targeted cell type. This is an example of a so-called "Trojan horse" particle.

Of course, there are a number of hurdles that must be overcome using this kind of approach, one of the largest of which is the fact that cancer cells are characterized by having a lot of actin and cytoskeletal material — this gives them increased surface roughness which makes them stick together and to nearby tissues, and this makes it difficult for drugs and nanoparticles to fully enter tumors. No doubt this will be tackled next. Perhaps the fact that cancer cells have reduced mechanical stiffness (i.e. they are softer) than normal cells — as determined using AFM, can be made use of in future therapies.

Nanoparticles — The Not so Good Ones

Although we mentioned this in Chapter 4, "Nanomaterials", we should look again briefly at nanoparticles, now that we are in bio-mode. Really what is all the fuss about? There are a number of key issues that we should address: Nanoparticles are small enough that they may inadvertently go places we don't want them to. This happens at several different levels associated with the body's defenses. First and foremost, our skin is the most important defensive barrier against foreign objects (Figure 26). As mentioned, when looking at cosmetics, small particles, be they liposomes or hard particles, can be small enough that they penetrate deeper into the skin than conventional pastes or creams.

If we look at the gross structure, we can see that there are a series of layers starting with the dead skin on the top, followed by the constantly regenerating cells (together this is called the "epidermis"). Below this (not shown here) is the dermis which contains blood vessels and sebaceous glands. Below this is

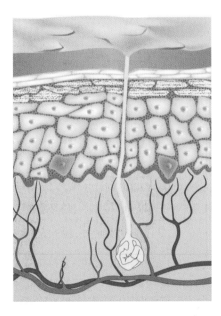

Figure 26. **Skin** — The body's first defensive barrier. Nanoparticles can travel between the cells to enter tissue and even blood vessels, from where they can travel through the body, but they have to get past the skin first.

the subcutaneous layer which contains mostly fat cells. The epidermis is the main barrier, which keeps moisture in and pathogens out. This top layer is often damaged, and has small flaws and holes, through which material can enter the body. If that material is a foreign object, then it can reside in the dermis for a long period of time and may act as a site where infection or cancer may start. On the whole, this is unlikely. The main issue is that skin cancer is triggered when the cells in the epidermis and below are damaged by UV light and start mutating, which is why sunscreen is so important. A secondary defense against this is the pigment, melanin, which is produced in the epidermis. Its function is to absorb UV light and prevent it from getting into the cells. Unfortunately, when melanin reacts with UV radiation, it transforms into oxymelanin which is transparent and ineffective. Thankfully, this triggers the melanin-producing cells in the epidermis to get to work producing more, but this does take time.

The second level of defense is blood vessels — by and large, they are relatively impermeable, unless they are weakened by any of a number of blood conditions. Very little gets through them, although they are of course permeable — otherwise how would oxygen, water, and nutrients get from our blood to our cells! Nanoparticles can of course seep through blood vessels, so once they enter the bloodstream they can in principle end up anywhere.

The third level of defense and perhaps the most important one, is the blood–brain barrier.

After all of these defenses have been breached, the next step is to get into our cells. This is usually a rather tricky job, as the cells are coated by a cell membrane, similar to the liposomes we saw earlier. This is a complex arrangement of lipids, containing a variety of channels through which materials can travel between the inside and outside of the cell (Figure 27). Gases such as O_2 and CO_2 can readily diffuse through the membrane, bur larger entities such as proteins and many ions (e.g. Na^+), need to travel through channels (illustrated in green) that open selectively when they recognize the entrant.

By and large, most nanoparticles are simply too large to pass through these channels, whose size is typically up to a few nm, so this is not something we particularly need to be concerned about in most cases. I say most cases as there is one situation that is problematic — single-walled carbon nanotubes (Figure 28). It has been shown that they are small enough in diameter (~1 nm) to fit through these channels and enter cells. It has also recently been shown that when the end of a carbon nanotube comes in contact with a cell, it becomes engulfed by the cell membrane and ends up partially inside as shown below.

From there, they can enter the nucleus, and as luck would have it, DNA has an affinity for carbon, so it attaches itself to the nanotubes, wraps itself around it like a climber, and then proceeds to mutate. For this reason, carbon nanotubes are

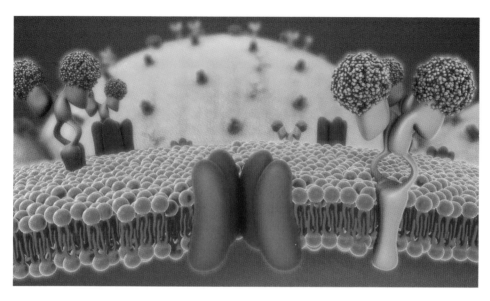

Figure 27. Cell membrane. This is a lipid bilayer. Gases can diffuse directly across the membrane, but proteins and ions can only travel through special channels — shown in green.

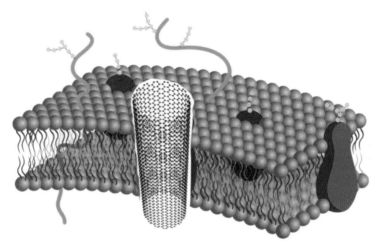

Figure 28. When single-walled carbon nanotubes encounter cells, they become engulfed by the cell membrane and end up penetrating the cell, where they come in contact with the nucleus.

considered to be as carcinogenic as asbestos, and I for one, will not handle them. Multi-walled carbon nanotubes are too large to do any of this and the evidence is that they are harmless, which is why they are still used in all manner of composite materials.

To be fair, many of us will not ever come in contact with carbon nanotubes, so while this may sound alarming, it's highly unlikely that it will become a problem.

A much more pressing problem unfortunately is the titanium dioxide issue. This is a substance that we have seen mentioned multiple times so far in the context of electronic devices, self-cleaning coatings on glass and of course, cosmetics. It has another use, which is in toothpaste and foodstuffs, where it is used for its whitening properties. It is a very strong white pigment (it is used in paints also, but not usually in nanoparticle form), and also has the property that it is an anti-caking agent, so it is found in very many foodstuffs including donuts, candies, and chocolate. It is used particularly in the kind of candies where there is chocolate surrounded by a colored sugary coating — it is the white substance between the two, and is one of the main ingredients of this coating. It is particularly prevalent in blue candies. Until 2015, it was used by Dunkin' donuts, where it was added to the powdered sugar coating to make it look brighter. At that point, they stopped using it due to pressure from consumer groups who pointed out its potential cancer and toxicity risks. The lethal dose of this substance is around 5 g per kg of body mass. Although this would amount to several thousand donuts, the risk of fatality is rather low. However, being serious, the titanium dioxide used is in nanoparticle form, and as it is taken orally in foodstuffs and toothpaste, it can move throughout the body and causes long-term damage. It has been shown that it can damage the visual cortex and cause inflammation of the hippocampus in the brain, it adversely affects male fertility, and interferes with the body's ability to fight infection. That's quite a rap sheet for just one material. To dig a little deeper into that last part, some recent work from the University of Cambridge has shown that when titanium nanoparticles are small enough (typically around or less than 8 nm in diameter), they can attach to T-cells (which are one of the body's main defense mechanisms), interfering with their normal operation. One of the key ways in which the immune system works is that if a pathogen is encountered, T-cells send molecular signals which lead to the formation of more T-cells in order to boost immunity. This leads to an inflammatory response, which can be a marked risk for cancer. It appears that titanium dioxide can sometimes give the same response as a pathogen and given that it is not broken down very efficiently by the body, it can persist for long periods of time, leading to a continuous level of inflammation. So, it is a double-edged sword, T-cells can see it as potentially harmful and take it up, but then it interferes with them. As to why T-cells have such a strong interaction with titanium dioxide is unknown, but as with most such things, it is either a shape effect or something related to the chemical nature of the titanium dioxide nanoparticle surface. This is an area of active research now and hopefully within the coming years, our understanding will deepen, and the learnings can be taken and applied to other areas of infection/inflammation. A rather beautiful if terrifying illustration of this problem is shown in Figure 29. This is an image of a section of the small intestine known as "Peyer's patch" — a region with a high degree of lymphatic

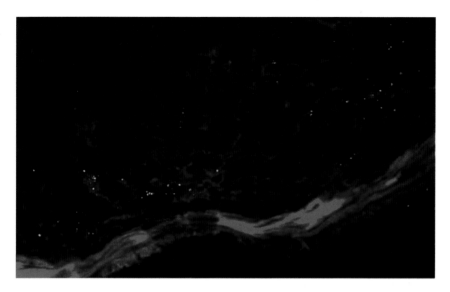

Figure 29.   Titanium dioxide nanoparticles (white dots) within the Peyer's patch of a section of the human small intestine. These build up over time and disrupt the immune system (*Image Courtesy*: Jonathan Powell & John Wills, Department of Veterinary Medicine, University of Cambridge).

tissue, which is central to immunity. These patches act in much the same way as the tonsils do in that they trap foreign objects, decide if they are harmful and if so, destroy them. The little white dots in the image are where titanium dioxide nano-particles (that were ingested) have become trapped. The experimental evidence is that as a person ages, they accumulate more of these particles, and ultimately this influences their ability to fight off infection. A sobering thought.

So, what is the bottom line as regards nanotechnology in healthcare is that there are positive and negative aspects. The negatives are that we do not yet fully understand the interactions between nanometer-sized objects and our cells and DNA, so we are all right to be cautious. The positives are that designer nanoparti-cles are leading us in the fight against cancer and other diseases, both in diagnos-tics and therapeutics/drug delivery. Appropriate ethical frameworks need to be established regarding these issues and many governments in the developed world (starting with the UK and the US) have already made significant advances in the right directions.

Chapter 8

Concluding Remarks

I hope that I have managed to convey some of the wonder and beauty of nanotechnology to you over the last seven chapters. I see nanotechnology as a natural development of the way we do science. The human race has come a very long way in a very short time. We can only hope that we have sufficient reason to be able to deal with our new technological capabilities with caution, both for ourselves, our children and the planet as a whole. It is only in the last 500 years that we started to apply scientific reasoning to describe the world around us. This boils down to evidence-based theories as opposed to religious, magical, or nice-sounding ones based on folk tales. This all happened so quickly that there is an understandable tension between the above viewpoints, but it does appear that the march toward the scientific approach is inevitable. Perhaps these are mutually exclusive and there does not need to be such tensions. Even deep science resorts on concepts that must sound magical or fanciful such as dark matter and pretty much anything in quantum mechanics. The human mind is simply not capable of comprehending all things, and it is ultimately our compassion and emotions that lead to the richness of the human experience, so they must not be overlooked.

We have seen that since we started to question the world around us rather than just accept things as *being*, we have developed a sophisticated set of theories about how things work, ranging from mechanics, materials, gravity, atomic forces, quantum mechanics, and electromagnetics. What we do not really know is *why* they work the way they do. As the theories became ever more detailed, experiments started to reveal inconsistencies, and this is really where the renaissance started at the turn of the 20th century. Relativity and quantum mechanics were developed within the course of a few years, and quantum mechanics in particular showed us that the world is indeed a very strange place if you look closely enough. With the insight of a few people including Feynman and Drexler, along with specific industrial needs, the tools at the heart of nanotechnology (scanning probe microscopes) were developed in the 1980s. Since then, we have transformed the way we do science, and the race is on to apply what we have learned in as many fields as possible. Some will bear fruit and others will not. We already make use of the novel properties of nanometer-sized things everyday, in the cosmetics we put on our faces, the microprocessors that enable our smartphones and computer devices to work, many polymers in sports equipment, and in medical diagnostics. The medical developments are of course the ones that will be of interest to most people, but bear in mind that they will also be the ones that take the longest to reach us — the timescale for medicines and therapies to go from the lab to the patient can be 20 years, whereas diagnostic tools are typically less than 10 years. We have very exciting times ahead, for sure. It is only through patience and continued targeted research that the promise of nanotechnology will come to fruition across the board, and we are at that crucial point right now where many of the initial promises have come to pass, and it is time for others.

Index